Intro
Freshv

by

Allan Pentecost

Rp

Richmond Publishing Co. Ltd.
Orchard Road, Richmond, Surrey, England.

First edition 1984

© Allan Pentecost 1984

ISBN 0 85546144 6 CASED
ISBN 0 85546143 8 PAPER
Published by The Richmond Publishing Co. Ltd.
Orchard Road, Richmond, Surrey, England.

Printed in Great Britain by
Kingprint Limited, Richmond, Surrey.

CONTENTS

Cover design: from a drawing by Allan Pentecost

Acknowledgements

I am indebted to Miss D. Britten, Prof. J.D. Dodge, Mr. F. Dobson, Dr. D.A. John, Dr. J.W.G. Lund F.R.S., Mrs J.A. Moore and Dr. C.M. Wood for advice and assistance with particular sections of this book and the many useful suggestions received. I am particularly grateful to Dr. Wood for reading the manuscript and the loan of a drawing. I also wish to thank staff at the British Museum (Natural History) for free access to the algal collections and Dr. B.S. Benedikz of Birmingham University Library for a print of Prof. G.S. West.

INTRODUCTION

People have been fascinated by freshwater algae since the first good microscopes became available in the early eighteenth century but the first British book on the subject was Dillwyn's *British Confervae* published in 1809. It contains many attractive plates of *Conferva*, a general name applied to all kinds of filamentous algae. One of the aims of the book was to "enliven the scenes of rural retirement" but few rallied to the call and it was another thirty years before Hassall's *History of Freshwater Algae* appeared, soon to be followed by Ralf's *British Desmids* in 1848. This was really the starting point for the serious study of freshwater algae in Britain and it coincided with the appearance of some important continental works. The next fifty years realized a veritable cornucopia of descriptive algology, originating mainly in Europe and the United States. It is only fairly recently that this descriptive 'tide' has swept other parts of the world but it is generally recognized that many algae are cosmopolitan so little descriptive work remains. What we do have is a vast collection of descriptive work, which, in many cases is proving a nightmare for modern taxonomists. The explosion of descriptive work in Britain was elegantly summarized in G. S. West's British Freshwater Algae of 1904. The book was updated by Prof. F. E. Fritsch in a second edition, "A treatise of Freshwater Algae" which is still a popular, extremely informative book although much outdated. Prof. G. S. West (Fig. 1) obtained an interest in algae from his father who was president of the Yorkshire Naturalists Union. He was an enthusiastic student of algae throughout his short life and also a skilled draughtsman. He was particularly interested in the desmids, a group of green algae, and published a beautifully illustrated desmid flora of Britain with his father between 1904-12. A new definitive algal flora of Britain is long overdue but unlikely to appear for a considerable time.

The author's own interest in the group began with the acquisition of a microscope in 1968 whilst an undergraduate. The

FIGURE 1

Professor G. S. West

instrument was purchased from the college for £6 and has seen continuous service ever since. It is still possible to buy such bargains from colleges and medical schools but this usually requires a personal contact in the profession. The microscope will always be the tool of the trade and many naturalists recognize the microscope as their most important and valued piece of equipment.

The algae are a diverse group of mostly aquatic plants. The freshwater forms tend to be small and the majority are microscopic although they often betray their presence by imparting colour to the water. When seen like this, most people find them unattractive, but they include some of the most beautiful microscopic objects of almost unending variety. Knowing this, it will not come as a surprise that most of the joys of collecting come at the end of the day, sitting with the microscope on the bench. Unlike most collectors, algologists have no idea of the day's haul until it is returned home, so it is useful to make a note of the collection sites in the field.

This book is designed to assist both amateur and professional algologists with the identification of the commoner freshwater algae and to bridge a gap between the shorter guides and the expensive reference works, many of which are in foreign languages or out of print. However, you should not be disappointed if you cannot identify algae to species level. Several thousand species of algae have been recorded from these islands, although it is far from certain that there are really quite so many morphologically distinct forms. Pure cultures of algae often show a considerable diversity of shape, so it is possible that some species are merely growth forms of others. Even if you fail to name the species, the generic discriptions are there to help you reach at least this level, and there is usually a guide to the number of species described together with the main characters used to distinguish between them. It is important to draw the plant, noting these characters and then proceed to a more detailed reference work if necessary.

EQUIPMENT

Excluding the microscope the amount of equipment required to collect and examine freshwater algae is quite small and inexpensive. Small glass bottles of 10-20 mls capacity, with press-on plastic lids are ideal for collecting (Fig. 2). This type rarely break and many can be carried without discomfort. Only small amounts of material should be added, as shown in the figure, otherwise the sample will deoxygenate and rapidly decay. Collections should be labelled in the field by writing in pencil on a small piece of paper and enclosing it in the bottle.

Planktonic − or free-floating algae are collected with a plankton net. This consists of a conical piece of bolting silk or nylon attached to a small cup at the cone apex and provided with a tap. The cone is kept rigid by a wire hoop to which a light rope is attached (Fig. 2). Various mesh sizes are available, but the holes should be 40-80µm (1 µm = 1/1000 millimetre) wide for most purposes. The net is either thrown in from the lakeside or towed slowly behind a boat. The plankton net will retain only the larger algae, and many small forms will pass through the net, so the samples obtained are rarely representative of the whole. Care should be taken not to disturb the sediment in shallow water and the net should always be washed after use. A sedimentation technique suitable for small algae is described on p. 53.

The most important item of equipment is of course the microscope. This should be a sturdy instrument fitted with x10, x40 and preferably (but not essential) a x100 oil immersion objective and a x10 eyepiece. Good microscopes are expensive, but suitable instruments can be purchased for the price of a good pair of binoculars. If you have no experience of microscope work, consult an expert and avoid the cheap instruments sold in toy shops.

Identification of algae relies a good deal on the measurement of the cells. For this purpose, an eyepiece graticule is required, which fits, or can be made to fit into the microscope eyepiece (Fig.

1

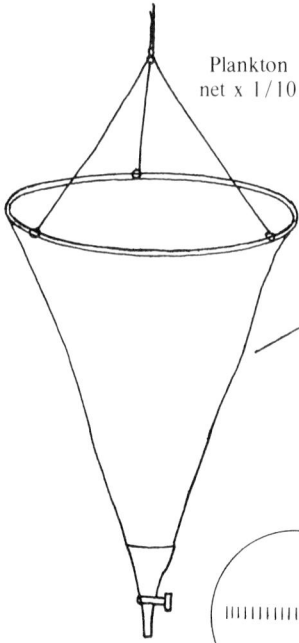

Plankton
net x 1/10

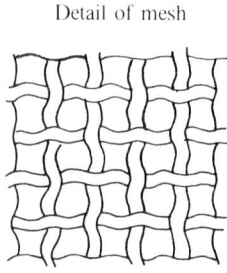

Detail of mesh

Specimen
bottle

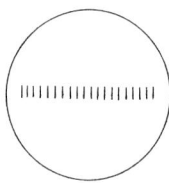

Graticule

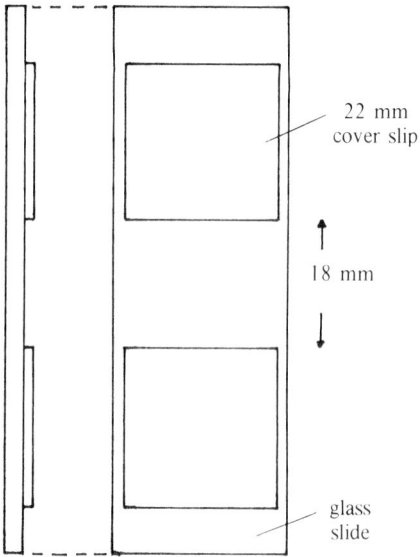

22 mm
cover slip

18 mm

glass
slide

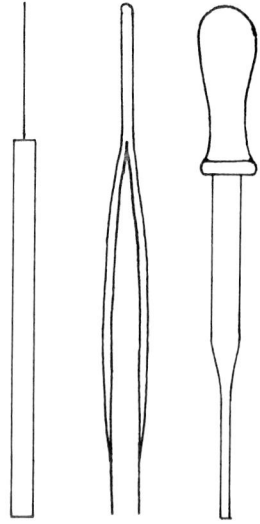

2). Graticules can be calibrated by focussing onto a finely ruled surface and this enables measurements to be made in microns (µm). These units are used throughout this book.

A mounted needle, a fine pair of forceps and a small dropper will be required to manipulate the specimens onto the microscope slide. Cavity slides (Fig. 2) should be made by sticking two coverslips onto a microscope slide with Araldite and curing at 70-100°C to give a good, even adhesion. When a coverslip is placed on top of the cavity, about 0.1 ml of sample can be examined. This is a far greater volume than can be studied by placing a coverslip directly onto the slide, but oil-immersion work is not feasible with a cavity.

Algae should first be examined live, using natural daylight illumination or a lamp fitted with a blue filter. Fixation will often be necessary, particularly for flagellates, which are hard to study whilst swimming rapidly. The best general fixative is Lugol's iodine, which is made by dissolving 2 g Potassium iodide and 1 g Iodine crystals in 300 mls water. This solution should be used sparingly. As well as fixing algae, it also reveals the presence of starch which stains a blue-black colour. Glutaraldehyde Solution (1%) is also useful but will not reveal the presence of starch. Much of the equipment mentioned here can be obtained from suppliers listed on p.226.

FIGURE 2 EQUIPMENT

EXAMINATION

Algae should be examined soon after collection, although it is possible to keep them reasonably fresh for up to two weeks in a refrigerator.

A single drop of water might contain several thousand algae belonging to numerous species. One algologist has claimed to have found over 200 species in a single drop of water but 5-15 species is about average in my experience, with a handful of really common types. The best way of recording collections is by means of accurate drawings or photographs. Samples can be preserved with a mixture of 1 part formalin, 3 parts alcohol and 6 parts water. After cleaning, diatoms can be prepared as permanent slides (see p.86).

Only a small quantity of material should be examined under the microscope, after having been teased out with needle and forceps if necessary. From the point of view of identification, perhaps the single most important feature of the algal cell, is the chloroplast. This is the body containing the chlorophyll and other photosynthetic pigments and often occupies about a third of the cell volume. Chloroplasts are lacking only in the cyanobacteria where instead the pigments appear to be distributed throughout the cytoplasm of the cell (Fig. 3a). The shape and colour of the chloroplast are important features in identification and wherever possible, a range of cells of a species should be examined since the shape tends to vary a little from cell to cell. The position of the chloroplast is also important. It is usually pushed up against the cell lining, next to the cell wall (parietal) but sometimes it occupies a central position, away from the wall (axile).

Some chloroplasts contain pyrenoids (Fig. 3b). These are spherical bodies 2-8 µm in diameter, composed of protein. They are usually obvious and occur in small numbers. Often there is a single pyrenoid centrally placed. Most green algae have pyrenoids which are surrounded by a starch sheath which stains with iodine but in other classes, the reaction is usually negative because other storage products are formed.

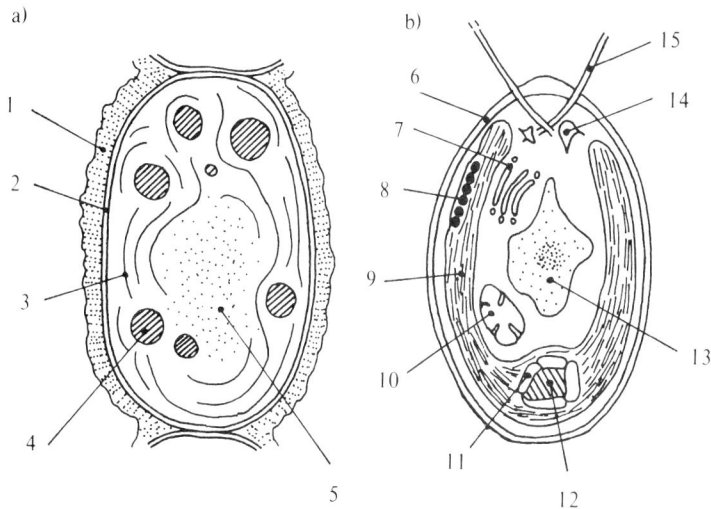

FIGURE 3

The fine structure of algal cells as seen
with the electron microscope

a)   *Anabeana* (cyanobacterium) x 4000
     1 sheath, 2 multilayered wall, 3 thylakoid, 4 cyanophycin (storage) granule, 5
     nuclear region.

b)   *Chlamydomonas* (Chlorophyceae) x 3000
     6 cellulose cell wall, 7 Golgi body, 8 stigma, 9 chloroplast with thylakoid
     stacks, 10 mitochondrion, 11 starch sheath, 12 pyrenoid, 13 nucleus, 14
     contractile vacuole, 15 flagellum.

In the flagellate algae and certain related groups, a small reddish 'eye-spot' or stigma may be present. In the green algae, the stigma is found within the chloroplast (Fig. 3b) but in other groups, such as the Euglenophyceae, it is found outside. The stigma is rarely more than 2 μm wide and normally close to the end of the cell containing the flagella. The *flagella* are fine hair-like structures whose movement gives motion to the whole organism. They beat up to 100 times a second, and being very narrow, (<0.5 μm,) they are difficult to see. Studies with the electron microscope have revealed a considerable range of flagellum types and this has aided algal classification. Some flagella propel the cell by a breast-stroke movement (most Green Algae) with the flagella in front of the cell whereas others produce sinuous movements and propel the cell from behind (Dinophyceae, Chrysophyceae). The range of flagellum types cannot easily be appreciated with the light microscope.

The nucleus, mitochondrion and Golgi system (Fig. 3b) although essential components of algal cells, excepting the cyanobacteria, are of no importance in identification and are usually invisible without special staining, but small pulsating spherical structures called *contractile vacuoles* are usually associated with flagella and these are sometimes important in classification.

It is now time to consider the plants as a whole and note their means of reproduction since these are also used in classification. Algae reproduce either sexually or asexually but the former is observed much less frequently than the latter. Most algae reproduce by simple fission to produce a pair of identical cells. This process may occur indefinitely, resulting in large aggregations or filaments of great length but in some filamentous species, limited growth occurs, this is particularly so in the branched forms. Another type of asexual reproduction also occurs and this is by means of *zoospores*. These are small flagellate cells produced by the subdivision of the contents of a single algal cell (Plate 35 Fig. 17) which may itself be motile. Each parent cell subdivides to produce 4-32 zoospores which are released through a pore or a break in the cell wall. In some algae, the zoospores are produced in special, enlarged *zoosporangia*. The zoospores swim away, then finally settle, lose their flagella and germinate to form new plants. Algae are the only green plants which produce zoospores, probably because these motile cells require a damp or aquatic environment. Many algae produce non-motile spores which are formed by a similar process. These are called *aplanospores. Autospores* are a type of aplanospore produced by many small algae, e.g. *Quadrigula* where the colonies develop in miniature before being released (Plate 32 Fig. 10).

Sexual reproduction is a fairly simple process in the algae and there is normally just one haploid life form in the freshwater forms, although in the Red Algae, there are often two distinct generations of plants. Sexual reproduction results in the combination of inherited characteristics from two haploid individuals followed by a redistribution and separation of the characteristics between the offspring. Cells which fuse resulting in combination are called *gametes*. These are sometimes produced in special cells called *gametangia* (Fig. 4b). The result of fertilization is a fusion cell which frequently secretes a thick, pigmented wall and may remain in a quiescent state for many months before separation and germination occur. Germination either results in the direct development of new plants or the production of zoospores.

There are three basic kinds of sexual expression in the algae.

a) Isogamy.

The most common method. The gametes are motile and morphologically identical (Fig. 4a) and cannot be distinguished as male or female. Fusion results in a zygote which germinates to release new individuals or zoospores. A few algae are *anisogamous* where the male and female cells differ in size but both are motile.

b) Oogamy.

Here only one gamete, the male, is motile and this swims towards the female which is larger and usually attached to another plant. The fusion cell is called an *oospore* in this group and is occasionally surrounded by other sterile cells (Fig. 4c). The oospore germinates to produce zoospores which establish new plants. Occasionally the plants are sexually distinct, e.g. some species of *Volvox, Oedogonium* and *Vaucheria.*

c) Conjugation.

Neither gamete is flagellate and the plants themselves must come into close contact. The walls of adjacent cells swell out and fuse together to allow the passage of the nuclei (Fig. 4b). The fusion cell, called the *zygospore*, germinates to produce 2-4 daughter cells and zoospores are not produced. This type of reproduction is seen in the Zygnematales (Green Algae) and most commonly observed in *Spirogyra.*

The reproductive process is used extensively in the classification of algae but great care must be taken in distinguishing between zoospores and motile gametes. In fact, zoospores often behave as gametes in isogamous algae so that the distinction then breaks down.

The cell walls of algae are composed of a variety of materials, but in the Green Algae it is mainly cellulose. These algae most closely resemble the larger land plants. Some flagellates possess an outer test or 'theca' which is often granular and pigmented, making the characters of the cell difficult to see. A good example is seen in *Trachelomonas* (Plate 13 Fig. 5).

The slippery feeling of algae is caused by the secretion of a jelly-like mucilage. Cells are often seen embedded in mucilage in a particular orientation and it is sometimes brightly coloured in colonial Blue-green Algae. It is best seen by adding a small drop of India Ink to the sample, as the ink particles cannot penetrate the material.

The algae of limestone areas are frequently encrusted with calcium carbonate and they may even bore into the rock itself. This can be removed by dissolving in 2.5% EDTA solution (ethylenediamine-tetraacetic acid) adjusted to pH 8. If this is unavailable, dilute hydrochloric acid usually will suffice.

FIGURE 4

Sexual reproduction in the algae

a) Isogamy in *Chlamydomonas*.
    I) loss of flagella and gamete formation, II) gametes liberated, III) fusion, IV) zygote, V) release of zoospores to produce new plants.

b) Oogamy in *Coleocheate*.
    I) male gametangium releasing gamete which fertilizes an oogonium, II) oospore surrounded by sterile cells, III) release of haploid zoospores.

c) Conjugation in *Closterium*
    I) recently divided cells, II) pairing, III) fusion resulting in zygospore formation, IV) zygospore germination producing two individuals.

a)

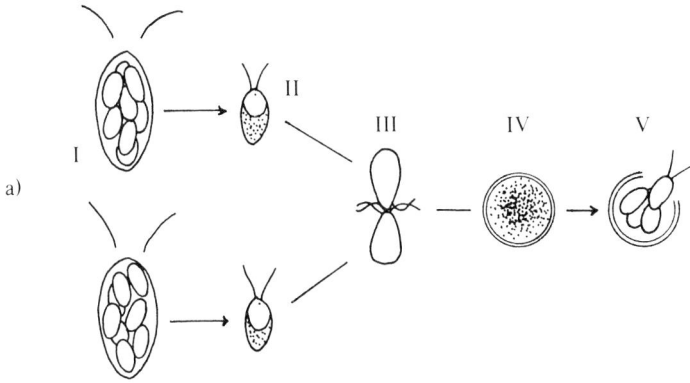

I II III IV V

b)

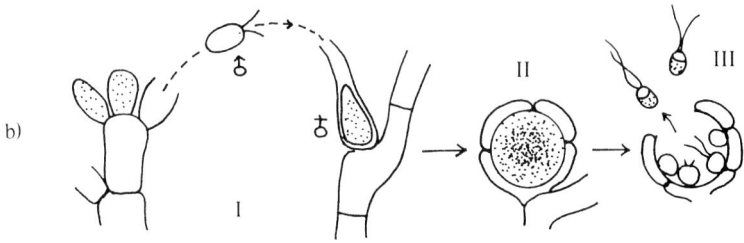

I II III

c)

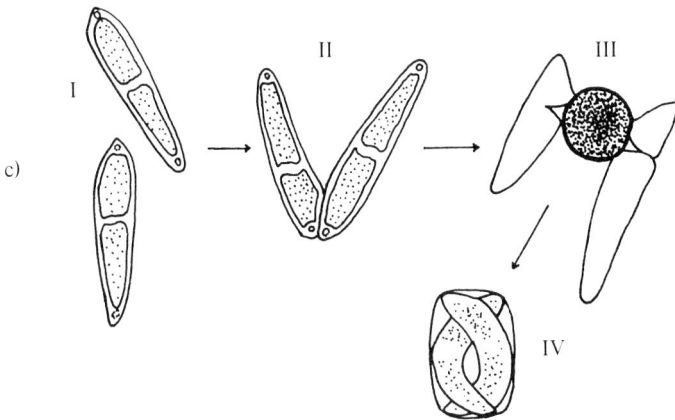

I II III IV

WHERE TO LOOK

Most algae are aquatic and occur either as plankton, or associated with surfaces (*benthic*) such as submerged plants (*epiphytes*), rocks (*epiliths*) or sediment (*epipelic*). A few species grow upon permanent snow or within the cell spaces of other plants. The habitats colonized are therefore diverse and also include the hair of sloths and polar bears, the backs of spiders, hot springs and chlorinated swimming baths.

The richest collecting grounds are boggy lake margins and ditches, especially where water plants such as *Utricularia* and *Callitriche* abound. The diversity of polluted waters tend to be low and normally consists of just a few species of cyanobacteria and green algae. Plankton is best collected from large lakes and slow flowing rivers. Fast flowing water has little plankton but the submerged rocks and plants are often richly covered with attached forms. These are best removed by scraping with a fingernail or by running the fingers down plant stems. Damp mosses are also a good source and should be squeezed out into a bottle, retaining a small moss sample for reference. *Sphagnum* moss harbours a rich variety of algae, the dark muddy patches found on hillsides in the west being a particularly good source.

Despite the multitude of algal species, an understanding of algal ecology has been obtained for merely a handful of freshwater species. A diversity of environmental and biological factors influence algal growth but of these, nutrient concentrations together with irradiance and temperature are the most important.

Water which contains an abundance of dissolved nutrients is capable of supporting a dense crop of algae. Nitrogen and phosphorus are essential to growth and if these are limited in quantity, little growth occurs. Water which is fertile, i.e. rich in dissolved nutrients, is termed *eutrophic.* In eutrophic situations, the deeper waters, which are often cool and dark, may become depleted of oxygen, particularly during summer. This results from the microbial breakdown of sedimenting algal and animal

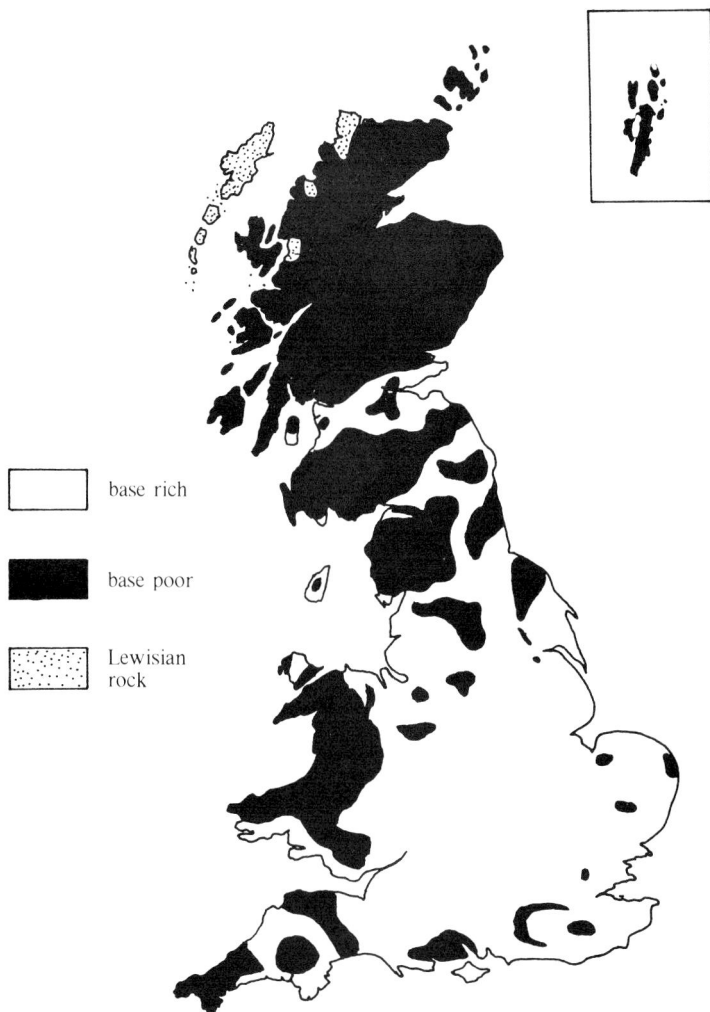

base rich

base poor

Lewisian
rock

FIGURE 5

Map showing predominantly base-rich and base-poor areas of Britain neglecting the
effects of organic pollution and fertilizers.

11

remains. As a result, fish, which need oxygen, are sometimes killed. The water may also begin to smell as a result of the decomposition. Some waters are naturally eutrophic, but many of our lakes and rivers have become so, through the input of treated sewage which contains high concentrations of nutrients. In contrast, nutrient-poor water supports little algal growth. This is evident in our British western mountain lakes which contain a solution which differs little from rainwater and is extremely clear and well oxygenated. This water type is termed *oligotrophic* and an intermediate where there is partial deoxygenation is called *mesotrophic*.

Generally speaking, water derived from catchments containing clays and silts tend to be meso- or eutrophic whereas those derived from sands, hard volcanic rocks and granites are oligotrophic as the nutrients are either released very slowly or present only in trace amounts. In Britain, most of the fine sediments occur in the south and east whilst the hard rocks are confined mainly to the north and west. The ancient Lewisian rocks of the far northwest appear to provide particularly rich sites for desmids, which include many of the most attractive algae. The distribution of base-rich, base-poor and Lewisian rocks is shown in Fig. 5 but no account has been taken here of the possible effects of water pollution by sewage or the effect of agricultural fertilizers, which can lead to eutrophication.

Table 1 indicates the kinds of algae which can be expected to occur in the types of water described above, but it should be considered as no more than a general guide. It can be seen that some species occur in all water types whereas others have a more restricted range. Table 2 lists, in decreasing frequency, the most common algae recorded in the author's notebooks from two aquatic habitats.

TABLE 1

OLIGOTROPHIC	MESOTROPHIC	EUTROPHIC

Asterionella formosa, Fragilaria crotonensis

Dinobryon divergens   *Anabaena, Aphanizomenon*

Oscillatoria (planktonic spp.)

Synura

Microcystis, Coelosphaerium

Volvox

Melosira granulata,
Stephanodiscus hantzschii

Zygogonium
ericetorum

Ceratium hirundinella

Cyclotella kuetzingiana

Botryococcus braunii

Cladophora spp.

Tabellaria flocculosa

TABLE 2

*Most frequently recorded genera of freshwater algae
taken from the author's lists.*

On and among submerged plants and stones	Lake plankton
Tabellaria	Dinobryon
Mougeotia	Asterionella
Trachelomonas	Coelosphaerium
Scenedesmus	Anabaena
Closterium	Ankistrodesmus
Cosmarium	Fragilaria
Spirogyra	Cryptomonas
Chlamydomonas	Oscillatoria
Synedra	Tabellaria
Fragilaria	Staurastrum
Cladophora	Volvox
Pediastrum	Ceratium
Cryptomonas	Microcystis
Euglena	Mallomonas
Oscillatoria	Eudorina
Zygnema	
Stigeoclonium	
Navicula	
Tribonema	
Cymbella	

KEYS TO THE MAJOR CLASSES

Notes on the keys and descriptions.

The key below should enable you to find which class an alga belongs to. Further information on the class will be found on the page cited, followed by a key to the main genera. Descriptions of the genera and selected species then follow. A few genera are accompanied by keys to the species.

Each genus description usually provides the following information: the range of cell and colony size, habit, ecology if known, details of common or important species; approximate number of species recognized worldwide, characters used to distinguish species, method of reproduction.

Abbreviations used in the descriptions.

$a > b$	a greater than b
$a < b$	a less than b
Ak	Akinete
Ap	Aplanospore
Aut	Autospore
d	diameter
DA	dorsal arc length (Desmids)
I	Isogamous
Id:	Key characters for identification of species; methods of reproduction; total number of species described (approx.).
ITM	Number of striations in 10 μm (Diatoms)
ltb	length to breadth ratio (length/breadth)
Syn.	Synonym.
μm	micrometre (1/1000 mm.)
z	zoospore

Key

1a   Cells without chloroplasts, with a diffuse blue-green, olive or red-brown colour, sheaths sometimes coloured. Cells with minute granules or obscured by dense granules caused by gas vacuoles in some planktonic forms. Cross wall generally, very thin <1 μm. No flagellates. **Cyanobacteria** p.20

1b	Cells with chloroplasts	2
2a	Chloroplasts pale to deep grass green	3
2b	Chloroplasts some other colour	4
3a	Starch test usually positive.	
3b	Starch test never positive	5

4a   Unicellular, colonial or filamentous normally with 1-2 parietal or axile chloroplasts, occasionally with many chloroplasts. Flagellates with two apical flagella of equal length ($\pm$ 10%), swimming in breast-stroke fashion. **Chlorophyceae** p129.

4b   Plants large and erect, typically 4-50 cm long with regular branches (c.f. horsetails), rooted by rhizoids in soft sediment. Cells large with numerous chloroplasts. **Charophyceae** p.222.

5a   Unicellular, colonial or filamentous, normally with two or more small discoid chloroplasts per cell. Flagellates with two apical flagella, one distinctly longer than the other ($>10\%$ difference). **Xanthophyceae** p63.

5b   Unicellular flagellates with one long, thick (to 0.5 μm) flagellum emerging from an apical depression. Chloroplasts 2-many, often discoid. Rods or discs of paramylum storage material conspicuous in most forms. Cell wall can appear elastic and finely striated. **Euglenophyceae** p75.

6a   Cells with a rigid, ornamented silica wall composed of two halves, boat-shaped, pill-box shaped or needle like, isolated or colonial, sometimes in filaments. Often motile but without flagella. Chloroplasts 1-several, brownish, shape variable. **Bacillariophyceae** (Diatoms) p83.

6b   Cells not so formed. 7

7a   Cells with two flagella lying partially within two deep furrows. One furrow girdling the cell, the other at right angles towards one apex. Cell wall smooth or composed of angular plates, cells spherical, often slightly flattened, sometimes with large projecting horns, solitary. Chloroplasts discoidal, numerous, brownish or rarely yellow, red or blue. Starch test often positive.

Dinophyceae p70.

7b   Cells without two deep furrows.   8

8a   Chloroplasts pale yellow to brown, usually 1-2 per cell. Cells single or colonial, rarely filamentous. Flagellate forms with either a single long flagellum or with one long and one short flagellum. Starch test negative. Spherical or urn-shaped silica cysts may occur. Mostly small flagellates.

Chrysophyceae p53.

8b   Chloroplasts some other colour, cysts absent, flagella not strongly unequal if present.   9

9a   Unicellular, bean-shaped flagellates with two slightly unequal flagella arising from a small depression or furrow. Chloroplasts 1-2, olive, red or blue, starch test usually positive.   **Cryptophyceae** p79.

9b   Plants filamentous and frequently with a complex structure, or encrusting, rarely solitary. Chloroplasts 1-several per cell, olive, red or blue. Starch test often positive. Usually attached to rocks and mosses in streams. No flagellates.   **Rhodophyceae** (Red Algae). p44.

Two other classes of freshwater algae are sometimes seen but neither is common. These are the Phaeophyceae, with one dark, encrusting genus, *Heribaudiella* (p.85) found on submerged sheltered rocks, often near the sea and the Chloromonadophyceae (p.84), a small class containing species with large cells, numerous green chloroplasts and two prominent flagella, one directed forwards and the other trailing.

THE CYANOBACTERIA OR
BLUE-GREEN ALGAE

This is an important group of organisms which resemble the bacteria far more closely than the true algae. The cells of the cyanobacteria have no chloroplasts, the chlorophyll-containing membranes, or *thylakoids* are spread throughout the cell (cf. Figs. 3a and b) and there are no nuclei or mitochodria. As a result, the pigments appear to be spread evenly throughout the cell. In addition to chlorophyll, two other pigments occur which make the cells appear olive or blue green. These are proteins forming red and blue complexes respectively. Similar pigments also occur in the Rhodophyceae, Cryptophyceae and Dinophyceae.

The cell walls of the cyanobacteria are very thin but often surrounded by a broad mucilaginous sheath which may be coloured, particularly in *Gloeocapsa*, *Scytonema* and *Stigonema*. Many of the filamentous forms show a gliding movement which seems to be connected in some way with the secretion of sheath material. Chains of cells are called *trichomes*, not filaments as is the case in other groups of algae. This is a result of the way the group has been classified in the past and it would be wise to continue its usage, despite the initial confusion it causes.

The cyanobacteria are a remarkable collection of organisms, capable of surviving under extreme conditions, e.g. in hot-springs, on desert rocks and below the Antarctic ice. Fossils closely resembling existing types have been found in rocks over two billion years old and they were probably largely responsible for our oxygen-enriched atmosphere. Today they are far less abundant and probably best known as 'nuisance organisms'. Dense growths or 'blooms' of *Microcystis* can block water treatment filters and also occasionally produce poisons which kill fish and cattle. Others produce strong musty odours which can taint water supplies and meat. Despite this, the cyanobacteria also have their uses. They are important primary producers and provide food for a wide variety of aquatic animals. *Spirulina* has been used as human food and contains a higher proportion of protein than soya bean. However, perhaps their greatest significance is as natural fertilizers. Many types can fix nitrogen gas into forms available for other plants such as rice. Some species of *Nostoc*, a nitrogen-fixing genus, form associations with fungi to produce lichens and they are also found in the roots of some cycads and flowering plants.

Some genera are distinguished by their ability to form *heterocysts*. These are pale, thick-walled cells unique to the cyanobacteria (Plate 5 Fig. 1) which are known to be sites of nitrogen fixation. Heterocysts are formed from ordinary cells and the number produced seems to depend upon the availability of combined nitrogen. When levels fall, heterocyst production increases.

The cyanobacteria have other unusual features. Many planktonic forms accummulate small packets of gas in their cells. These *gas vacuoles* act as buoyancy regulators and occur mainly in *Anabaena*, *Coelosphaerium*, *Oscillatoria* and *Microcystis*. The vacuoles scatter light strongly and the cells appear dark and granular under the microscope as a result (Plate 6 Fig. 7). These genera frequently produce thick scums at the water surface in eutrophic lakes and reservoirs. If a dense sample is decanted into a stout glass container enclosed by a rubber bung with the air excluded, the cells will form a thick upper layer. Striking the bung sharply with a small hammer collapses the gas vacuoles resulting in a dramatic change of colour followed by the sinking of the cells.

There is no true sexual reproduction in the cyanobacteria but asexual spores are often produced. These consist of enlarged vegetative cells enclosed by a thickened, often brown wall. They are produced by some filamentous forms and frequently develop next to a heterocyst (Plate 5 Fig. 1). Some of the coccoid species produce *baeocytes*. These are minute spherical cells produced by internal cleavage and are often produced in large numbers. *Chamaesiphon* produces similar cells by budding (see Plate 1 Figs. 1-3). Many filamentous species are not known to produce spores. Instead, scattered cells appear to be killed off deliberately so that the trichome is broken into short lengths. These fragments are called *hormogonia* and are usually motile (Plate 4 Fig. 4). In *Scytonema* and related genera, the outer sheath is tough and remains intact whilst the trichome within breaks up. As a result, the broken ends of the trichome burst through the sheath and 'false branching' results (Plate 3 Fig. 5). True branching occurs in a few genera but it is usually erratic and irregular. Dichotomous branching, which is frequent in the Green Algae is rare and confined to exotic genera.

The cyanobacteria have been traditionally classified as true plants, not bacteria and it is only fairly recently that their affinities with the bacteria have been widely recognised. This is a serious problem as far as naming is concerned because bacteria are classified in an entirely different way to plants. For the person wishing to identify material in the field, rather than cultured strains, the traditional approach would appear to be most appropriate, in the absence of any practical alternative. The descriptions given in this book have been based upon this system, which relies heavily upon morphology but is not too far removed from recent bacteriological classifications. No attempt has been made to enumerate species and the group is badly in need of revision since some characters, e.g. sheath structure which is used widely to define genera are now known to be unreliable.

The cells of cyanobacteria average around 5 µm in diameter and therefore tend to be smaller than those of the algae which average around 12-22 µm.

Cyanobacteria Key to Common Genera

1a   Cells solitary or in colonies, reproducing by fission, budding or baeocytes. Heterocysts and spores absent.   2

1b   Cells forming simple or branched trichomes. Heterocysts and spores may be present.   16

2a   Cells elongate, club- or urn-shaped, encrusting rocks or plants, reproducing by baeocytes or budding.   3

2b   Cells spherical to shortly cylindrical, sometimes gas-vacuolate, planktonic or encrusting rocks, reproducing by simple fission or baeocytes.   5

| 3a | Cells elongate to club-shaped, reproducing by budding. |
| | *Chamaesiphon* 7 |

| 3b | Reproducing by baeocytes | 4 |

| 4a | Cells with an acute apex terminated by a hair (rare). |
| | *Clastidium* 8 |

| 4b | Cells rounded or urn-shaped, no hair. | *Dermocarpa* 10 |

| 5a | Cells solitary, spherical to shortly cylindrical, dividing in one plane. | *Synechococcus* 1 |

| 5b | Cells in colonies | 6 |

| 6a | Colonies consisting of a single layer of cells arranged in a square lattice. | *Merismopedia* 2 |

| 6b | Colonies otherwise | 7 |

| 7a | On rocks or plants or growing within limestone or shell. 8 |

| 7b | Planktonic or on fine sediments, often with gas vacuoles 13 |

| 8a | Growing within rock or shell, cells in irregular rows resembling trichomes. | *Hyella* |

| 8b | Growing upon rocks or plants | 9 |

| 9a | Colonies dark, cartilaginous and cushion-like. | 10 |

| 9b | Colonies pale and jelly-like or dark with the cells forming a thin non-cartilaginous stratum. | 11 |

| 10a | Cells in tiers forming colonies up to 2.5 mm. d. often coalescent, no baeocytes. | *Hydrococcus* |

| 10b | Cells arranged irregularly, reproduction by baeocytes only. |
| | *Dermocarpa* |

| 11a | Cells spheroidal to fusiform, ltb 1.5-3, in soft mucilaginous colonies dividing in one plane. | *Coccochloris* 3 |

| 12a | Cells in small packets surrounded by sheaths which are frequently laminate and coloured, dividing in 2-3 planes. |
| | *Gloeocapsa* 4 |

12b Cells in large irregular colonies not divided into small packets. *Microcystis* 5

13a Cells elongate-spheroidal or fusiform, in loose mucilaginous colonies, ltb 1.5-3. *Coccochloris* 3

13b Cells spherical, spheroidal or tear-drop shaped, ltb <1.5, often gas-vacuolate. 14

14a Cells joined together in small mucilaginous packets, mucilage sheaths often lamellate and occasionally tinted, normally hyaline in planktonic forms, no gas vacuoles *Gloeocapsa* 4

14b Colonies large, consisting of many cells, mucilage diffluent and non-lamellate. 15

15a Cells closely packed into spherical colonies up to 300 μm in diameter. *Coelosphaerium* 6

15b Cells densely or loosely arranged in large, (0.5-3 mm) irregular colonies. *Microcystis* 5

16a Filaments unbranched or with false-branching only (Nostocales). 17

16b Filaments with true branching, sometimes with occasional false-branching (Stigonematales). 30

17 Filaments attenuated to hair-like points (Rivulariaceae) 20

Filaments not attenuated. 18

18 False branching absent. 19

False branches present (Scytonemaceae) 23

19 Heterocysts present (Nostocaceae) 25

Heterocysts absent (Oscillatoriaceae) 29

20 Heterocysts absent, filaments unbranched. *Homeothrix* 14

Heterocysts present, sometimes with false-branching. 21

21 Filaments united into cushion-like colonies, 1-30 mm in diameter, attached or free-floating. 22

Filaments solitary, often surrounded by a brown sheath.
Calothrix 15

22 Planktonic with spores adjacent to the terminal heterocyst, often gas-vacuolate. *Gloeotrichia* 16

Attached to stones or aquatic plants, gas vacuoles and spores absent. *Rivularia* 17

23 Heterocysts absent. *Plectonema* 18

Heterocysts present. 24

24 All false branches single. *Tolypothrix* 19

Some false branches arising in pairs. *Scytonema* 20

25 Heterocysts terminal with adjacent spores.
Cylindrospermum 21

Heterocysts intercalary or if terminal then without adjacent spores. 26

26 Filaments aggregated into parallel bundles forming floating rafts resembling specks of grass. *Aphanizomenon* 22

Filaments not forming rafts. 27

27 Filaments surrounded by broad gelatinous envelopes producing large (1 mm − 20 cm) spherical or spreading thalli over soil or rock. *Nostoc* 23

Filaments forming minute non-mucilaginous colonies, planktonic, (often gas-vacuolate.) or terrestrial. 28

28 Planktonic algae with spherical, or spheroidal cells.
Anabaena 24

Aquatic or terrestrial algae with markedly flattened cells, ltb 0.2-0.6. *Nodularia* 25

29 Filaments helically twisted, abundant in tropical alkaline lakes, less common elsewhere. *Spirulina* 12

Filaments straight or irregularly twisted, forming dense mats or dispersed in the plankton, often motile.
Oscillatoria 13

30   Thalli gelatinous and spreading, with pedicellate heterocysts, (rare outside the tropics).   *Nostochopsis* 26

       Thalli cartilaginous or gelatinous, heterocysts intercalary.
          31

31   Filaments multiseriate with short uniseriate side branches, thalli dark, cartilaginous and irregular.   *Stigonema* 27

       Filaments uniseriate with irregular side branches coming off at right angles.   *Hapalosiphon* 28

Order Chroococcales

1. Synechococcus

Cells solitary, elongate-spheroidal to cylindrical, 2-10 ltb, sometimes with a narrow sheath. The species occur frequently both in the plankton and on rocks and soil. One species is common in hot springs. The cells range from 2-8(15) μm long. (Id: cell ltb). (Syn. *Anacystis*.) There are two widespread British forms.

    Cells 12-15 μm long, 2-3 ltb.   *S. aeruginosus*
              (Plate 1 Fig. 6).

    Cells 4-7 μm long, 2-7 ltb.   *S. elongatus* (Plate 1 Fig. 5).

2. Merismopedia

Cells spherical or spheroidal, forming a single-layered stratum, sometimes gas vacuolate and measuring 1-5(8) μm long. There are several common forms which occur on lake sediments, among aquatic plants and in the plankton, two of which are described and figured. (Id: cell ltb, cell separation).

    Cells 5-8 μm wide, closely packed.   *M. elegans*
              (Plate 1 Fig. 9)

    Cells 3-4 μm wide, separated by mucilage. *M. punctata*
              (Plate 1 Fig. 8)

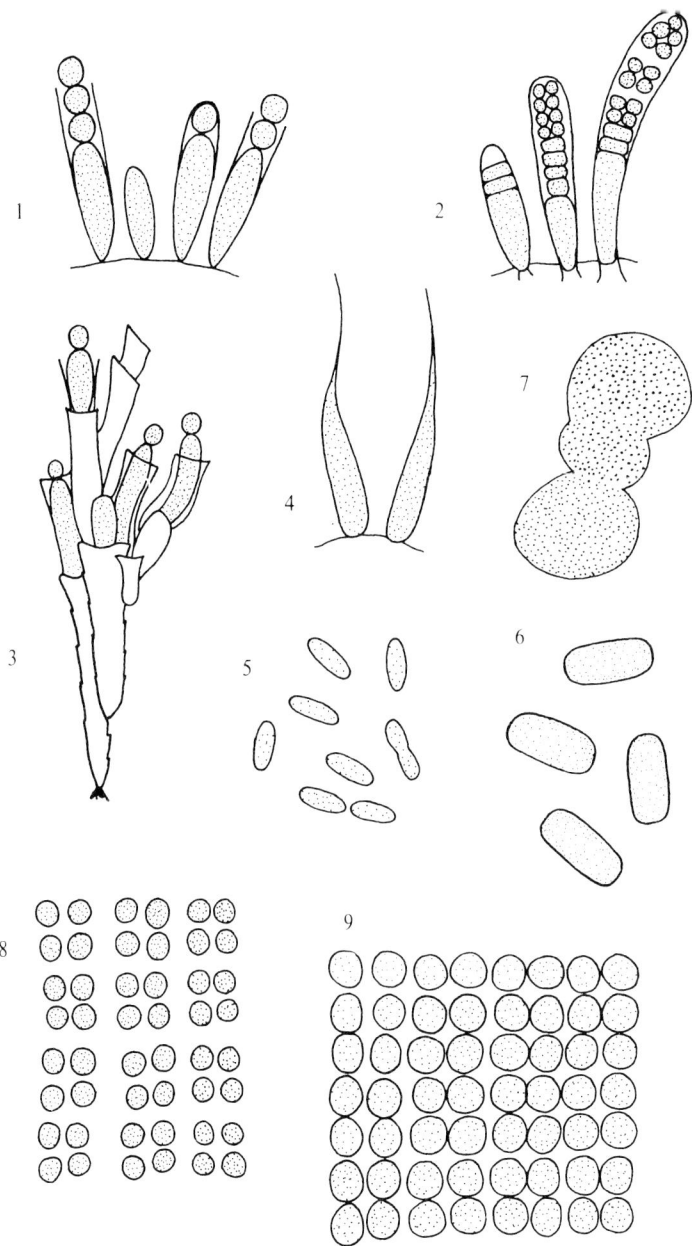

PLATE 1

1 *Chamaesiphon incrustans* x 1000, 2 *C. confervicola* x 1000, 3 *C. fuscus* x 1000, 4 *Clastidium setifegum* x 1800, 5 *Synechococcus elongatus* x 1600, 6 *S. aeruginosus* x 2200, 7 *Microcystis wesenbergii* colony x 125, 8 *Merismopedia punctata* x 1000, 9 *M. elegans* x 750

25

3. Coccochloris

Cells spheroidal to elongate-cylindrical, 3-9 (15) μm long, ltb 1.5-3, in loose mucilaginous colonies, pale blue-green. The species are usually found on damp rocks (Id: habitat, colony size, Syns. *Gloeothece, Aphanothece*).

1   Cells 7-9 μm long, colonies up to 2 mm in diameter.
C. stagnina (Plate 2 Fig. 5)

1   Cells 2-6 μm long, colonies small, usually less than 1 mm.
C. peniocystis (Plate 2 Fig. 4)

4. Gloeocapsa

Cells enclosed in distinct, often lamellate sheaths 1-15 μm in thickness which are colourless or tinted, forming small to large, irregular colonies in the plankton or on rocks. Cells excluding sheath (0.3)2-10(40) μm wide. The species are distinguished mainly on the cell size and sheath colour. They are both variable and care must be taken to ensure that the cells are not mounted in acid or alkaline media as this can affect the sheath colour. The species are abundant throughout the world, particularly on steep rocks subject to periodical inundation where they may form coloured streaks or 'Tintenstriche', best seen in Britain at Malham Cove in Yorkshire.

Key to the common British species

1a   Sheath colourless   2

1b   Sheath coloured, growing on rocks   3

2a   Cells 8-32 μm, sheaths up to 10 μm, often lamellate, in bog pools among *Sphagnum*.   *G. turgida* (Plate 4 Fig. 7)

PLATE 2

1 *Hyella fontana*, Gordale, Yorks x 1600, 2 *Hydrococcus rivularis*, Blaen Pennant, Gwynedd x 1700, 3 *Dermocarpa aquae-dulcis* x 650, 4 *Coccochloris peniocystis*, Malham, Yorks x 900, 5 *C. stagnina*, Bangor, Gwynedd x 900, 6 *Gloeocapsa punctata* x 750, 7 *G. rupestris*, Malham, Yorks x 800, 8 *G. sanguinea*, Malham, Yorks x 800, 9 *Microcystis aeruginosa* colony x 60, 10 *M. flos-aquae* colony x 60

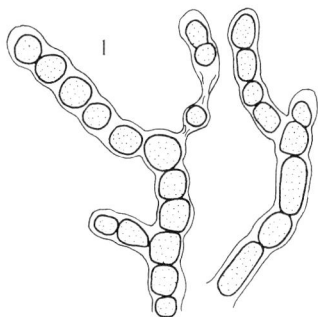

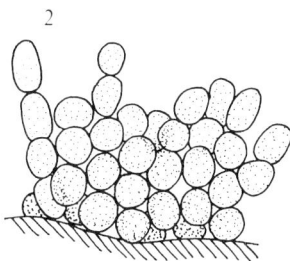

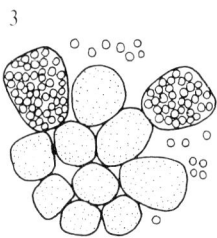

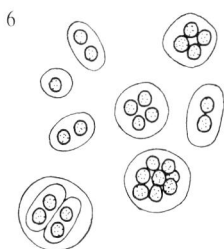

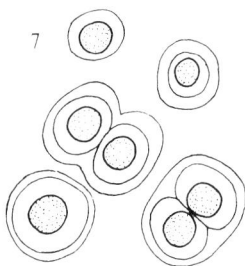

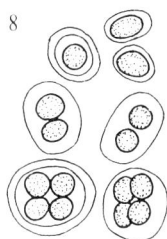

| 2b | Cells 0.7-3 µm, sheaths 2-15 µm, non lamellate, frequent on sheltered rocks. *G. punctata* (Plate 2 Fig. 6) |

| 3a | Sheaths yellow-brown. | 4 |

| 3b | Sheaths red or blue. | 5 |

| 4a | Cells (5)6-8(11) µm, sheaths lamellate, 2-8 µm, colonies to 200 µm, common. *G. rupestris* (Plate 2 Fig. 7) |

| 4b | Cells (2)3-4(5) µm, sheaths 2-5 µm, sometimes lamellate, common. *G. kuetzingiana* |

| 5a | Sheaths reddish, 2-12 µm thick, variable in colour and lamellation. Cells (3)4-8 µm, common and variable. *G. sanguinea* (Plate 2 Fig. 8) |

| 5b | Sheaths blue-violet, 2-8 µm, lamellate, cells 4-6 µm, in small colonies, frequent on limestone. *G. alpina* |

5. Microcystis

Cells spherical or spheroidal, pale blue-green or dark and usually gas vacuolate, in large, often irregular mucilaginous colonies. Some species form extensive water blooms in eutrophic lakes, others occur on damp rocks and hot springs. Many species have been described on the basis of the mucilage shape and structure. Cells (1)2-7(9) µm in diameter, colonies 40 µm – 3 mm.

(Id: colony size and shape, Syn. *Aphanocapsa*).

| 1a | Colonies growing on damp rocks, 1-10 mm diameter, cells 3-6 µm wide, no gas vacuoles. *M. grevillei* |

| 1b | Planktonic, cells often gas vacuolate | 2 |

| 2a | Mucilage well defined at colony edge, colonies lobed and perforated, cells separated from each other in centre of the colony. *M. wesenbergii* (Plate 1 Fig. 7) |

| 2b | Mucilage diffluent and indistinct. | 3a |

| 3a | Colonies elongate and perforated, cells (3)4-6(9.4) µm wide. (Plate 2 Fig. 9) *M. aeruginosa* |

3b   Colonies irregular but rarely elongate and perforated, cells
      4-6 µm wide.   *M. flos-aquae* (Plate 2 Fig. 10)

6. Coelosphaerium

Colonies roughly spherical with spherical, spheroidal or tear-drop shaped cells 2-7 µm long. Common planktonic algae in eutrophic lakes. The cells are usually packed with gas vacuoles. (Id: ltb).

1a   Cells spherical, 2-5 µm diameter, colonies 40-100 µm. *C. kuetzingianum* (Plate 3 Fig. 2)

1b   Cells spheroidal, 5-7 µm long, ltb 3-7, colonies 50-300
      µm.   *C. naegelianum* (Plate 3 Fig. 1)

Order Chamaesiphonales

7. Chamaesiphon

Encrusting unicellular or colonial algae with elongate-clavate cells producing spherical 'exospores' at their apices. Cells 10-50 (100) µm long, 2-7 µm wide, often bent, exospores 2-5 µm, produced in one or several rows. Sheath usually present, often tinted yellow-brown, occasionally divergently lamellate. The species grow on rocks or aquatic plants, usually in clean, fast-flowing water. (Id: ltb, sheath structure, exospore number).

1a   Cells solitary and narrowly clavate.   2

1b   Cells forming small dendroid colonies, 5-20(30) µm long,
      ltb 5-8, sheath often divergently lamellate and brownish.
      Exospores 1-2, uniseriate.   *C. fuscus* (Plate 1 Fig. 3)

2a   Exospores 1-3, uniseriate, cells often arcuate, 7-30 µm
      long, ltb 5-8.   *C. incrustans* (Plate 1 Fig. 1)

2b   Exospores 8-15, in multiseriate rows, cells 15-40 µm long,
      ltb 8-12.   *C. confervicola* (Plate 1 Fig. 2)

8. Clastidium

Cells solitary or in groups and terminated by a long hair.

C. setifegum has cells 8-30 µm in length, ltb 5-8, producing 3-5 uniseriate endospores. It is rarely seen epiphytic on other algae or water plants (Plate 1 Fig. 4).

9. Hydrococcus

Thalli blackish-brown, encrusting, 20-200 µm thick, consisting of close-packed tiers of cells (2)3-6 µm wide.

The only species is *H. rivularis* which grows on hard stones in clean running water, mainly in the west (Plate 2 Fig. 2).

10. Dermocarpa

Thalli dark and encrusting but without a tier-like arrangement of cells. Reproduction is by baeocytes, 0.5-2.5 µm in diameter. Cells spherical, spheroidal or club-shaped, 8-25 µm wide (Id: cell shape). The species occur uncommonly as epiphytes or on rocks in streams.

D. aquae-dulcis (Plate 2 Fig. 3) has weakly clavate cells 15-20 µm long releasing numerous baeocytes.

11. Hyella

The colonies are small with an irregular, almost filamentous structure up to 200 µm long, penetrating calcareous rocks and old shells. The cells are pale blue-green and 10-30 µm long, ltb 1-8 with a wide sheath. There are few species.

H. fontana has cells 4-8 µm wide and is frequent in some parts of Britain growing within the Carboniferous Limestone (Plate 2 Fig. 1).

PLATE 3

1 *Coelosphaerium naegelianum*, Windermere x 400, 2 *C. kuetzingianum*, x 650, 3 *Gloeotrichia natans*, colony x 20 and trichome with heterocyst and adjacent spore x 375, Malham Tarn, 4 *Calothrix parietina*, three trichomes. x 500, 5 *Plectonema tomasianum*, Malham, Yorks, x 350

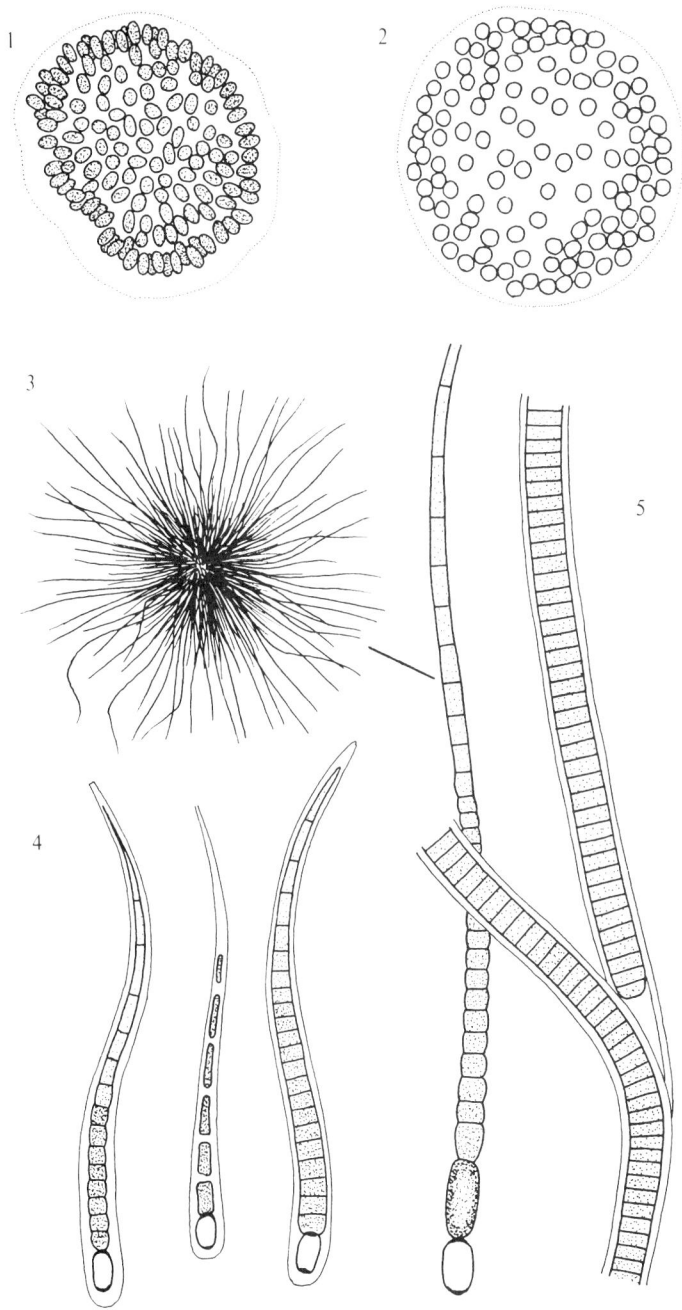

Order Nostocales

12. Spirulina

Cells helically twisted like a corkscrew, often motile and 0.5-10 μm wide, frequently gas vacuolate. The cross walls are often obscure under the microscope. The degree of trichome twisting has given rise to the description of numerous species, but studies with cultures have shown this to be an unreliable character. (Id: helix shape, gas vacuoles).

S. subsalsa is widely distributed although uncommon in Britain wth trichomes 0.4-6 μm wide (Plate 6 Fig. 3). It is usually found in stagnant water. *S. platensis* has trichomes 6-8 μm wide and is a well known plankton alga of alkaline tropical lakes.

13. Oscillatoria

Trichomes straight or bent, often of great length and either isolated or in dense bundles. Sheaths are absent and gliding motility is common in the benthic forms. Trichomes olive, blue-green or brownish, (0.5)-1.5-20 (60) μm wide. Gas vacuoles are frequent in the planktonic forms and the cells are disc-like, not barrel-shaped as in *Anabaena*. There are several closely related genera which are mentioned briefly below. (Id: shape of end cell, habitat).

Key to illustrated species

1a   Planktonic and normally gas vacuolate, common in eutrophic lakes.   2

1b   Benthic over mud, stones or plants, without gas vacuoles. 3

2   Trichomes 0.8-2 μm wide, cell ltb 2-5 gas vacuoles often polar.   *O. redeckei* (Plate 6 Fig. 8)

      Trichomes 4-6(9) μm wide, cell ltb 0.8-2, sometimes in loose bundles.   *O. agardhii* (Plate 6 Fig. 7)

3a   Trichomes (15)20-35(60) μm wide, with rounded, sometimes thickened end cells. Cell narrow and disc-like, widespread but uncommon.   *O. princeps*

3b   Trichomes (3)4-6(8) µm wide with narrowed and hooked end cells which may be thickened. Common mat-former avoiding bright light.   *O. brevis* (Plate 6 Fig. 5)

Related Genera

In *Microcoleus* the trichomes are enclosed within a common sheath and are sometimes twisted in a rope-like fashion. There are two widespread benthic forms which produce black leathery strata, often associated with organic pollution. *M. lyngbyaceus* (Plate 4 Fig. 4) has trichomes 5-15 µm wide with numerous small granules extending along the walls. *M. vaginatus* is similar but with granules only along the cross walls and trichomes 4-7 µm wide.

The trichomes are also enclosed within a common sheath in *Schizothrix* but here the cells are smaller and granules are lacking. *S. calcicola* (Plate 6 Fig. 6) is a common species in limestone areas where it forms broad yellow-brown strata on wet rocks. The trichomes are 0.8-4 µm wide and the sheaths sometimes trap particles of calcium carbonate resulting in the formation of small nodules of tufa. These may coalesce and form massive deposits such as those at Gordale in Yorkshire.

Lyngbya often occurs as isolated trichomes. It is similar to *Oscillatoria* but a firm sheath is present, enclosing just one trichome. *L. martensiana* (Plate 7 Fig. 4) has trichomes 5-10 µm wide and thick, unstratified sheaths.

Pseudanabaena combines the characters of an *Oscillatoria* and an *Anabaena*. It forms chains of unsheathed, barrel-shaped cells 1-10 µm wide but heterocysts are lacking. *P. minuta* (Plate 6 Fig. 4) has narrow trichomes (1.5-3 µm) and sometimes occurs in stagnant water.

14. Homeothrix

Trichomes 4-15 µm in diameter at their bases and gradually attenuated to a narrow hair. Heterocysts absent but often with a fine hyaline sheath. The species generally grow as a thin turf on submerged plants and stones in hard water (Id: sheath structure).

H. janthina is a small species with the trichomes 1-3 µm wide at the base and a thin sheath, often growing in dense tufts about 1 mm high on stones in hard water. Plate 4 Fig. 6.

15. Calothrix

Trichomes tapered with a basal heterocyst and usually a broad and often lamellate sheath, 5-10 μm wide at the base and (0.05)0.2-2(4) mm in length, occurring as isolated tufts or forming a thin turf on damp calcareous rocks, or submerged in hardwater streams and lakes. (Id: spore and heterocyst position, sheath structure)

C. parietina is probably the best known species with a yellow-brown, usually lamellate sheath up to 5 μm thick and trichomes (4)6-10 μm wide at their base (Plate 3 Fig. 4). The species sometimes intergrade with small colonies of *Rivularia*. Spores are rare.

16. Gloeotrichia

Colonies usually free-floating in hard water lakes, 0.2-4(10) mm in diameter, consisting of radiating and attenuated trichomes 7-10 μm in diameter at their base. Cylindrical spores 30-60(300) μm long. occur adjacent to the basal heterocysts, ltb 4-10).

Cells gas vacuolate, planktonic with spores 40-80 μm in length, trichomes 8-10 μm wide at their base.
G. echinulata

Cells normally without gas vacuoles, spores 50-250 μm long, attached to stones or plants, sometimes planktonic, trichomes 7-9 μm wide at their base. *G. natans* (Plate 3 Fig. 3)

PLATE 4

1 *Anabaena flos-aquae,* Windermere x 580, 2 *Tolypothrix tenuis* x 500, 3 *Hapalosiphon laminosus* x 800, 4 *Microcoleus lyngbyus,* Nanjing, China x 550, 5 *Rivularia biasolettiana,* Malham, Yorks, trichomes x 250, sectioned colony x 1, 6 *Homeothrix janthina* x 1500, 7 *Gloeocapsa turgida* x 400

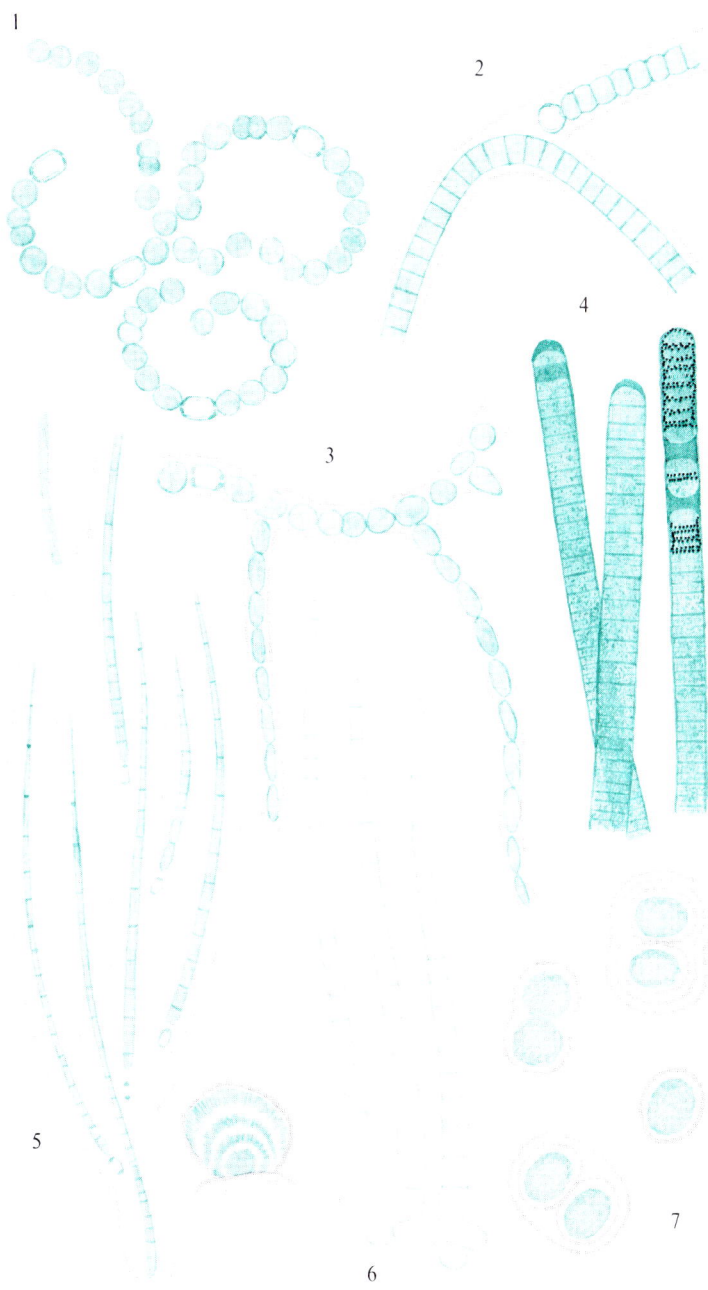

1

2

4

3

5

6

7

17. Rivularia

Thalli forming dark green or brown shiny cushions 2-20 mm in diameter, often coalescing to form strata and encrusted with calcium carbonate. Trichomes frequently false branched, 4-9 µm wide at their base, spores absent. If the thalli are cut away, a regular zonation pattern is often apparent and this can be used to date the plants. (Id: colony structure, degree of encrustation).

R. dura is a species which forms densely encrusted brownish or blue-green colonies on rocks and plants in hard water lakes and streams. The trichomes are 4-9 µm wide at their base and they are surrounded by wide, colourless or brownish sheaths (Plate 4 Fig. 5).

18. Plectonema

Trichomes false-branched, without heterocysts, 1-20(50) µm in diameter. Forming thin, dark strata over damp soil and rocks in streams (Id: sheath development, cell ltb).

P. tomasianum is the best known species with trichomes 11-25 µm wide with a lamellate hyaline or brownish sheath up to 3 µm thick (Plate 3 Fig. 5). It is most frequently seen as strata on damp rocks.

19. Tolypothrix

Trichomes false-branched, 4-10(16) µm wide with a hyaline or brownish sheath. Heterocysts present, intercalary or terminal at the sites of false branching, occasionally several in a row. These algae usually grow on damp rocks, soil or in small streams (Id: sheath thickness ans structure, cell ltb).

T. tenuis is a common species with trichomes 4-10 µm wide and a pale or brown lamellate sheath 1-3 µm thick. (Plate 4 Fig. 2).

PLATE 5

1 *Cylindrospermum stagnale*, R. Wey, Surrey x 950, 2 *Aphanizomenon flos-aquae* x 400, 3 *Nostoc piscinale* x 640, 4 *N. microscopicum*, Arncliffe, Yorks x 400, 5 *Scytonema myochrous* Gordale, Yorks. x 400, 6 *Nodularia spumigena*, Speldhurst, Kent x 630, 7 *Scytonema hofmanii* x 340

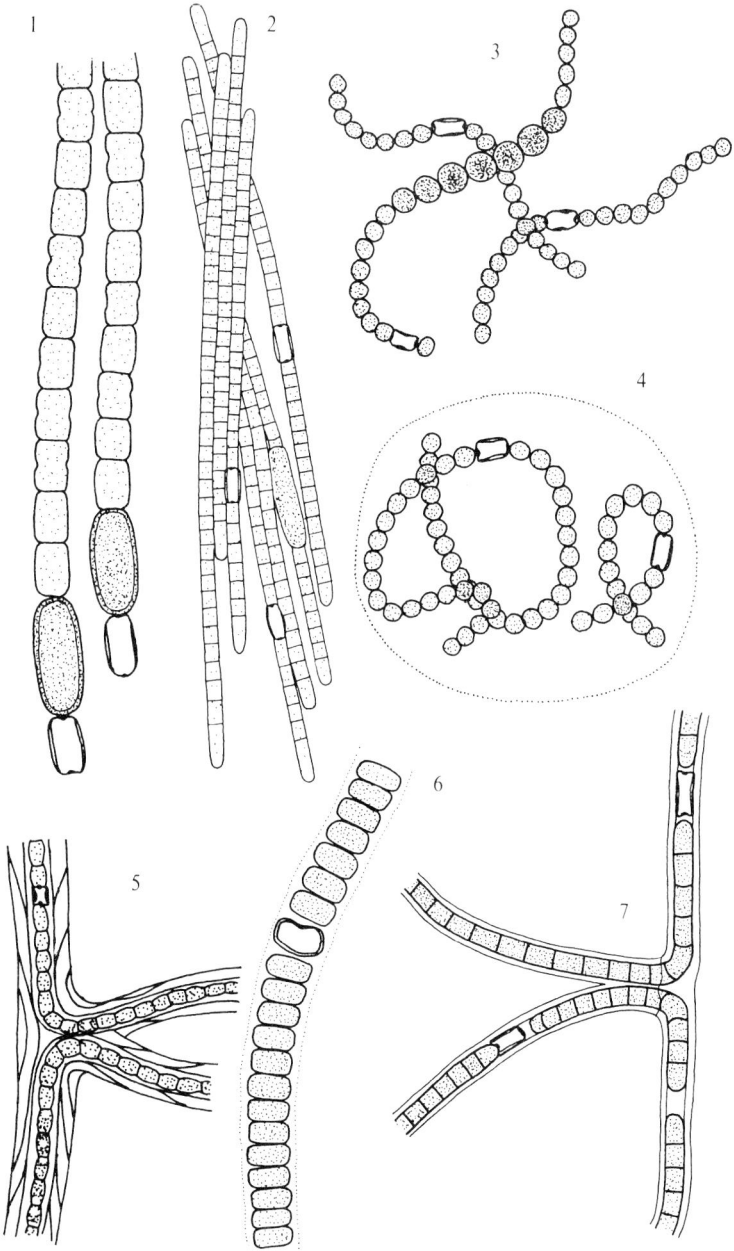

37

20. Scytonema

Scytonema is similar to *Tolypothrix* but the false branches usually occur in pairs giving a distinctive appearance. The trichomes are 3-10(16) µm in diameter with a thick, frequently lamellate and brownish sheath. Heterocysts terminal and intercalary. The species occur on damp rocks, soil and in small streams. (Id: sheath thickness and structure, cell ltb).

> Sheath divergently lamellate, 3-8 µm thick, trichomes 6-12 µm wide.   *S. myochrous* (Plate 5 Fig. 5)

> Sheath not lamellate, 2-3 µm thick, trichomes 5-10 µm wide.   *S. hofmanii* (Plate 5 Fig. 7)

21. Cylindrospermum

Trichomes (3)4-6 µm in diameter, unbranched, cells barrel-shaped, growing in thin dark green strata. Spores brown and thick walled, 20-30(40) µm long, ltb 1.5-5, adjacent to the terminal heterocysts (Id: spore wall structure, spore shape).

C. stagnale has smooth cylindrical spores 30-40 µm long, LTB 3-4 and trichomes 4-5 µm wide, and is a frequent species of damp soil and stagnant ponds (Plate 5 Fig. 1).

22. Aphanizomenon

Trichomes aggregated into parallel greenish bundles 1-4 mm in length. Cells 5-6 µm wide, the terminal cells usually longer, ltb up to 10. Spores cylindrical, hyaline, 60-80 µm long.

A. flos-aquae is a common bloom-forming species of eutrophic lakes with trichomes 5-6 µm wide and intercalary heterocysts and spores. Gas vacuoles are usually present but both spores and heterocysts can be difficult to find (Plate 5 Fig. 2).

23. Nostoc

Trichomes consisiting of chains of barrel-shaped or spherical blue-green cells within a broad mucilaginous envelope forming shiny green or brown colonies on damp calcareous rocks and soil, occasionally in streams and lakes. Thalli less than 1 mm to several cm. in diameter. Trichomes 3-5(7) µm wide with intercalary heterocysts. Spores pale or brownish, 6-12 µm long, ltb 1-3. *Nostoc* is probably the most commonly collected alga and has been used as food in the Arctic. (Id: colony size and shape, habit, spore and cell shape).

Key to common species

1a   Spores spheroidal, ltb 2-3, trichomes 4-8 µm wide, on soil or damp rocks.   2

1b   Spores spherical and hyaline, 6-7 µm in diameter, usually in chains. Cells spherical or barrel-shaped, 3-7 µm in diameter. Filaments more or less twisted in irregular colonies on damp rock or soil.   *N. piscinale* (Plate 5 Fig. 3)

2a   Thallus large and membranaceous, up to 20 cm.
                                               N. commune

2b   Thallus small and spherical, 1-10 mm in diameter.
                     N. microscopicum (Plate 5 Fig. 4)

24. Anabaena

Trichomes straight or coiled and twisted, often aggregated into small irregular bundles. Cells spherical, spheroidal or barrel-shaped, broad mucilaginous envelope lacking, (2)3-6(10) µm in diameter. Spores spherical or spheroidal, hyaline to brownish, (7)10-20(35) µm long, ltb 1-3. Heterocysts intercalary, often terminal in disintegrating material. Common bloom-forming algae, usually with gas vacuoles. (Id: spore shape and size, degree of trichome coiling).

Key to common species

1a   Trichomes straight or irregularly coiled   2

1b   Trichomes 6-8 µm wide coiled in an open helix.
                         A. spiroides (Plate 7 Fig. 3)

2a   Spores cylindrical with truncated ends, often in chains, 8-14 µm long, ltb 1.3-1.6, cells barrel-shaped 4-6 µm wide, trichomes straight or twisted.
                     A. variabilis (Plate 6 Fig. 2)

2b   Spores cylindrical or spheroidal, without truncated ends.   3

3a   Trichomes straight, 4-6 µm wide, spores 20-40 µm long, ltb 3-4, pale or brownish.   *A. oscillariodes* (Plate 6 Fig. 1)

3b   Trichomes coiled and often aggregated, 4-8 µm in diameter, spores 20-40 µm long, ltb 4-7, often bent, pale or yellowish.   *A. flos-aquae* (Plate 4 Fig. 1)

25. Nodularia

Trichomes composed of discoid cells usually constricted at the cross walls, 6-12 µm wide, ltb 0.2-7-0.6. Heterocysts intercalary, usually slightly flattened, sometimes with short yellow-brown spores.

N. spumigena has trichomes 8-12 µm in diameter and usually occurs as greenish strata on soil. The spores are brownish and flattened, 12 µm in diameter. Some species are planktonic gas vacuoles (Plate 5 Fig. 6).

Order Stigonematales

26. Nostochopsis

Trichomes irregularly branched, branches nearly perpendicular to the main axes, cells (1)2-10(15) µm in diameter with prominent lateral and terminal heterocysts. Thalli gelatinous, dark green and often lobate growing on damp rock, soil or in lakes or rivers. Frequent only in the tropics.

N. lobatus has barrel-shaped cells 4-10 µm in diameter with the trichome apices usually slightly tapered. Heterocysts lateral or on greatly reduced side branches, 10-12 µm long, ltb 1-1.5 (Plate 7 Fig. 2).

PLATE 6

1 *Anabaena oscillariodes* x 950, 2 *A. variabilis* x 750, 3 *Spirulina subsalsa* x 2000, 4 *Pseudanabaena minuta*, Tunbridge Wells, Kent. x 2000, 2 *A. variabilis* x 750, 3 *Spirulina subsalsa* x 2000, 4 *Pseudanabaena minuta*, Tunbridge Wells, Kent. x 2000, 5 *Oscillatoria brevis*, R. Medway, Kent x 650, 6 *Schizothrix calcicola*, Guilin, China. x 1250, 7 *Oscillatoria agardhii* with gas vacuoles x 830, 8 *O. redeckei*, Lough Neagh x 1600

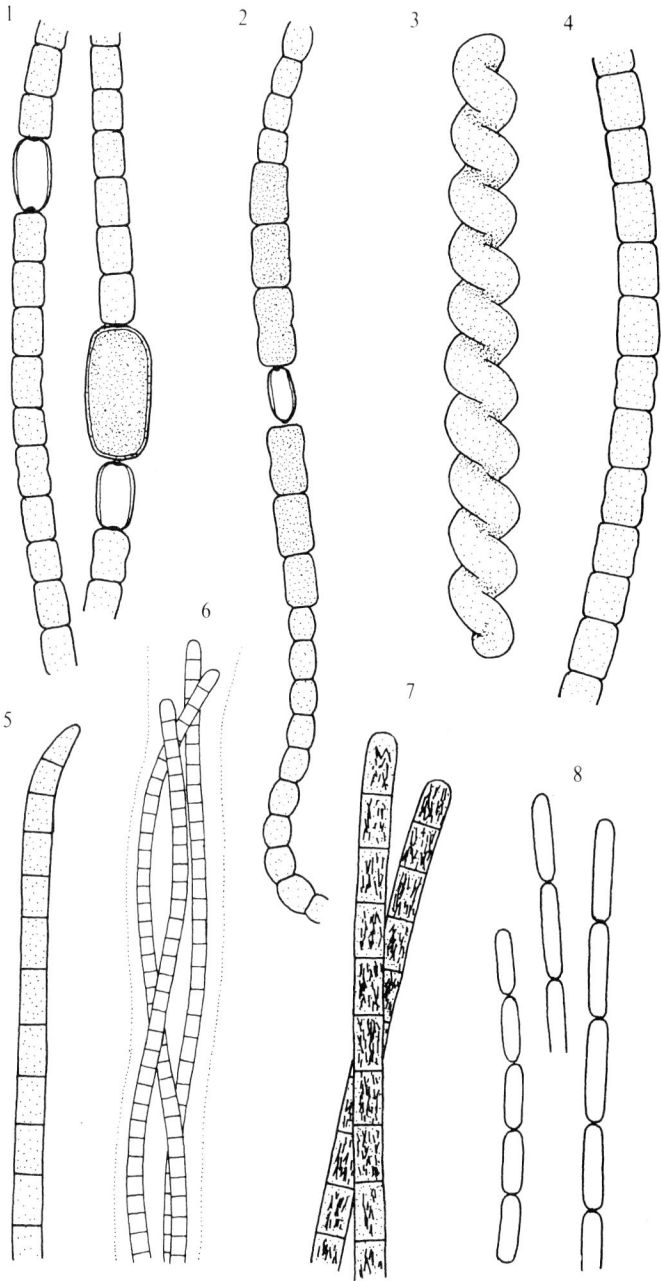

1

2

3

4

6

5

7

8

41

27. Stigonema

Thalli dark brown, forming a thin turf 1-5(12) mm thick over soil, rocks or trees. Main axes of the multiseriate filaments 15-40(70) μm wide. Side branches usually perpendicular to the main axes and uniseriate, often functioning as hormogonia. Cells 8-14 μm in diameter, frequently in groups of 2 or 4, with occasional scattered heterocysts. Sheath, yellow-brown in colour. (Id: thallus structure, size and shape).

> Filaments delicate, main axes biseriate, 30-40 μm wide with long, irregular side branches.
>
> *S. ocellatum* (Plate 7 Fig. 5)

> Filaments robust, multiseriate 40-60 μm wide with irregular main axes and short side branches.
>
> *S. mamillosum* (Plate 7 Fig. 1)

28. Hapalosiphon

Trichomes with irregular, often long side branches about the same width or slightly narrower than the main axes. Branches usually perpendicular, sometimes Y-shaped, heterocysts intercalary, spores rare, cylindrical, 5-12 μm long. Cells spheroidal or barrel-shaped, pale blue-green, (3)4-6(9) μm wide, usually enclosed by a hyaline sheath (Id: trichome structure, sheath colour).

H. fontinalis occurs in base-poor lakes and ponds, often among vegetation, with well branched trichomes 12-24 μm wide. The side branches are slightly narrower and the prominent sheath is pale or yellowish.

Mastigocladus laminosus is very similar and a well known hot spring alga which thrives at 30-63°C often forming dense filamentous flocs in the water. The trichomes are 4-6(8) μm wide and the side branches are sometimes poorly developed or even lacking. It is cosmopolitan (Plate 4 Fig. 3).

Both species should probably be referred to the genus *Fischerella*.

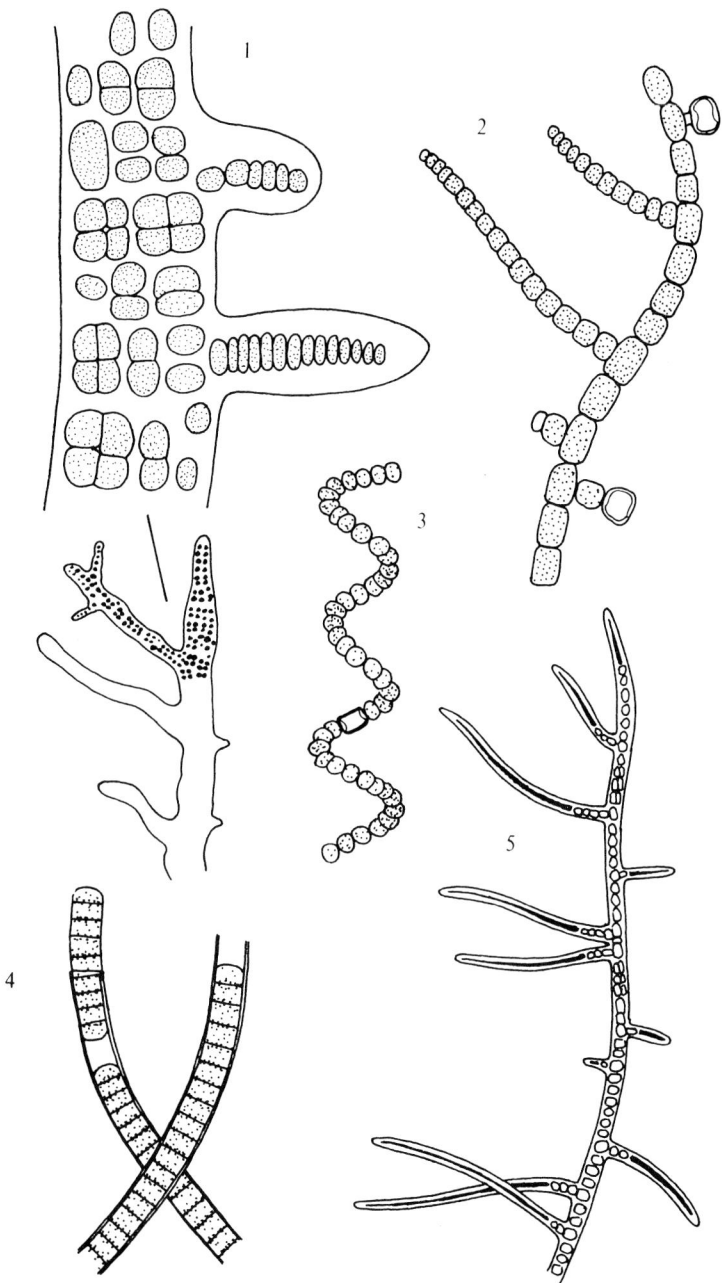

PLATE 7

1 *Stigonema mamillosum*, R. Duddon, Cumbria plant x 60, filament x 550, 2 *Nostochopsis lobatus* x 600, 3 *Anabaena spiroides* x 350, 4 *Lyngbya martensiana* x 600, 5 *Stigonema ocellatum* x 80

THE RHODOPHYCEAE OR RED ALGAE

This is a small but well defined freshwater group although only a few species actually appear red in colour. Most are dull olive-brown or blackish. The cells contain one to several chloroplasts which are ribbon-shaped or axile and stellate. Pyrenoids are frequent and the storage products are similar to starch giving a dark brown reaction with iodine. The large forms generally have a complex structure which either consists of a single, prominent main axis frequently surrounded by narrow cortical cells (uniaxial structure) or an axis composed of a large number of diverging threads resembling a fountain (multiaxial structure).

Some species have complex life cycles and reproduction processes. The sex organs are usually borne on the apices of side branches. The female cell or *carpogonium* is sedentary and equipped with a narrow apex or *trichogyne* towards which the male cells, or *spermatia* are attracted. The spermatia are remarkable in lacking any obvious means of movement and consist of minute spherical cells. Once attracted to the female, they pass down the trichogyne and fertilize the egg which then proceeds to divide and form a small plant attached to the parent. This plant eventually produces diploid *carpospores* which are released and establish new plants. To confuse the situation further, the new plants do not necessarily resemble the parent. For example, in *Batrachosperum* the carpospores give rise to simple filamentous *Chantransia*-stages (Plate 8 Fig. 4) which proceed to give rise to mature *Batrachosperum* plants. In some species, there are three morphologically distinct plants all representing the same species. Asexual monospores are also produced in some species (Plate 8 Fig. 4).

PLATE 8

1 *Hildenbrandtia rivularis*, R. Chess, Middlesex, colony xl, section x 400, 2 *Heribaudiella fluviatilis* section x 400, 3 *Porphyridium purpureum*, Rome, x 1400, 4 *Chantransia* stage of *Batrachospermum* with monospores, x 350, 5 *Bangia atropurpurea*, Queen Mary Reservoir, London, x 400, 6 *Batrachospermum vagum* with carpogonia, x 250, 7 *B. atrum* x 25

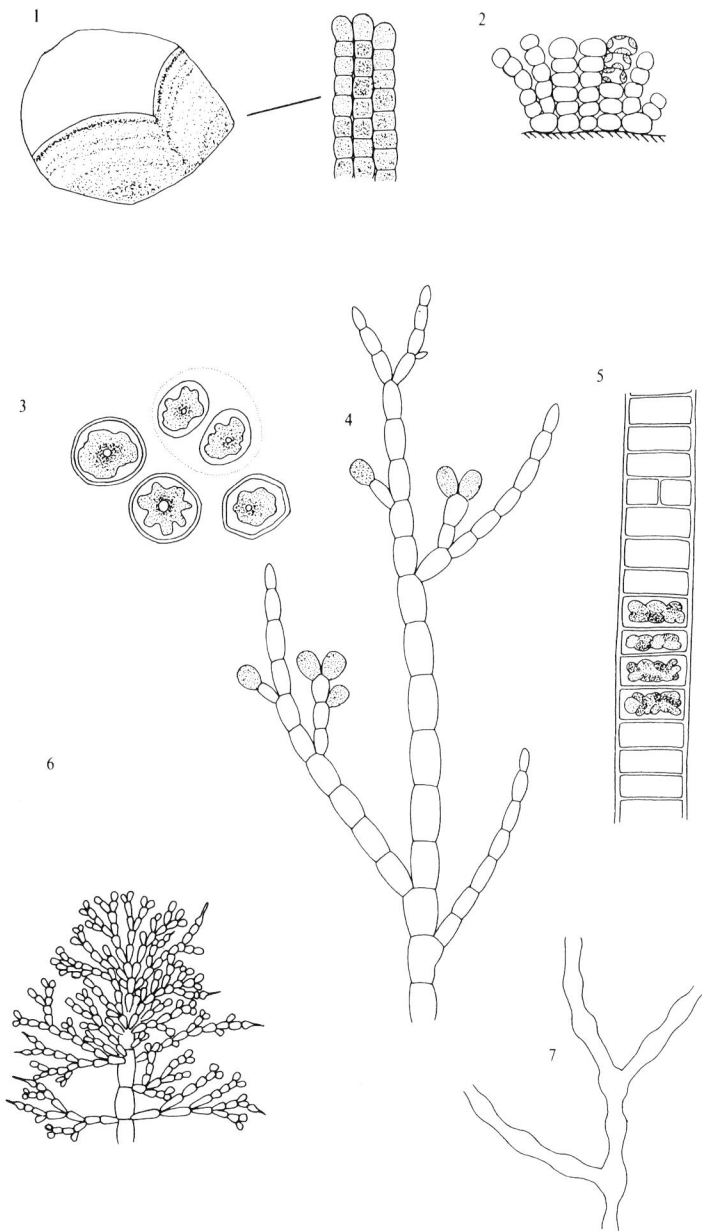

1

2

3

4

5

6

7

45

Key to genera

1a   Unicellular and terrestrial *or* multicellular and encrusting rocks in streams, reddish in colour.   2

1b   Filamentous, often large and olive in colour.   3

2a   Cells spherical or angular with an irregular axile chloroplast and pyrenoid, mostly on damp calcareous substrata.   *Porphyridium* 1

2b   Encrusting hard stones forming blood-red or brown patches, cells in close-packed tiers.   *Hildenbrandtia* 2

3a   Microscopic simple or occasionally branched filaments   4

3b   Large forms, uni- or multiaxial, (0.5)2-40(200) cm.   5

4a   Filaments unbranched, cells with a lobed, reddish-brown chloroplast and a pyrenoid.   *Bangia* 3

4b   Filaments well branched, cells with several reddish, discoid chloroplasts usually without pyrenoids, ltb 1.5-2.5, often associated with *Batrachosperum* or *Lemanea*. Filaments 10-14 µm wide, spherical monospores frequent. *Chantransia* stages (Plate 8 Fig. 4).

5a   Thalli lax and mucilaginous with regular tufts of branches developing from a main axis which is often corticate.   *Batrachospermum* 4

5b   Thalli without regular tufts of branches   6

6a   Thalli bushy or bristle-like with a series of regular annular thickenings   7

6b   Thalli lax or stiff, without annular thickenings   9

7a   Thalli when sectioned revealing a loose spongy layer around the central axis, bristle-like, rarely bushy. *Lemanea* 5

7b   Thalli solid when sectioned, irregularly branched or bushy (non British)   8

8a   Central axis surrounded by a layer or rounded cells

enclosed by close-packed radiating filaments (S. hemisphere)   *Nothocladus* 6

8b   Central axis surrounded only by branched, radiating filaments (U.S.A.)   *Tuomeya* 7

9a   Thalli multiaxial, developing at the edge into a soft felt of loose filaments. Lax and mucilaginous, usually branched.
                                                          Thorea 8

9b   Thalli uniaxial, lax and filiform or cartilaginous, olive to brown-black, rare in Britain.   *Compsopogon* 9

1. Porphyridium

Cells 6-12 µm wide forming large pink or red colonies on damp walls and rocks.

P. purpureum is the commonest species and is widespread on calcareous material, particularly near the sea. (Plate 8 Fig. 3).

2. Hildenbrandtia

H. rivularis, is frequently seen growing on stones in calcareous streams, particularly on flint or igneous rock. The thalli are 1-10 cm in diameter and the cells which are 7-12 µm wide, form close packed vertical tiers 50-200 µm high (Plate 8 Fig. 1).

3. Bangia

B. atropupurea (Plate 8 Fig. 5) forms long dark red filaments (15)20-40(60) µm wide, with a few cells sometimes forming lateral divisions. The cells are mostly short and discoid and the species is sometimes found attached to mosses and rocks in lowland streams, often near the sea.

4. Batrachospermum

These are the well-known 'frog-spawn' algae, so called because of their abundant mucilage. The thalli are very soft, olive brown or grey-blue in colour and are usually found in clean flowing water in the shade, frequently close to springs.

The thalli are composed of a main axis which is often slightly branched, with regular tufts of branches composed of barrel shaped cells. The carpogonia will be found at the apices of the tufted branches. Once fertilized, these develop into dense globular masses of cells (Plate 9 Fig. 1) which give rise to carpospores. The species are difficult to identify, a major character being the shape of the trichogyne (Plate 9 Figs. 2a-e), although there are three widely distributed forms recognized in Britain. The thalli range from (1)5-20(40) cm in length.

1a   Lateral branches greatly reduced in length.   *B. atrum*
            (Plate 8 Fig. 7).

1b   Lateral branches well developed   2

2a   Tufted appearance little evident, branches developed uniformly along the axis.   *B. vagum* (Plate 8 Fig. 6)

2b   Branches in distinct, regular tufts, with urn-shaped trichogynes.   *B. moniliforme* (Plate 9 Fig. 1)

5. Lemanea

Thalli olive, brown or blackish, consisting of simple or branched tapered bristles, (1)4-12(20) cm. long and (0.3)0.5-1.5(2.5) mm wide. Dark patches of male cells can often be seen on the annular thickenings. Two types of structure are found which form the basis of the classification and four species have been reported in Britain, occurring in fast flowing water, mainly in hilly districts. (Id: colour, shape of male cell patches, internal structure; 17).

1a   Central axis clothed in fine filaments, thalli usually unbranched. Sugenus   *Eulemanea*Plate 10 Fig. 1c). Thalli olive-brown, 5-8 cm long, up to 2 mm wide. Patches of male cells in a narrow broken ring.   *L. torulosa*
            (Plate 10 Fig. 3).

PLATE 9

1 *Batrachospermum moniliforme*, R. Wandle, London, showing habit x 7.5, part of axis with cell clusters around fertilized carpogonia x 80 and antheridia x 500, 2 Carpogonia of a) *B. vagum*, b) *B. moniliforme*, c) *B. arcuatum*, d) *B. virgatum* and e) *B. boryanum*. x 500, 3 *Lemanea mamillosa* x 2, 4 *L. fluviatilis* habit x 1.5, detail of antheridial area x 6

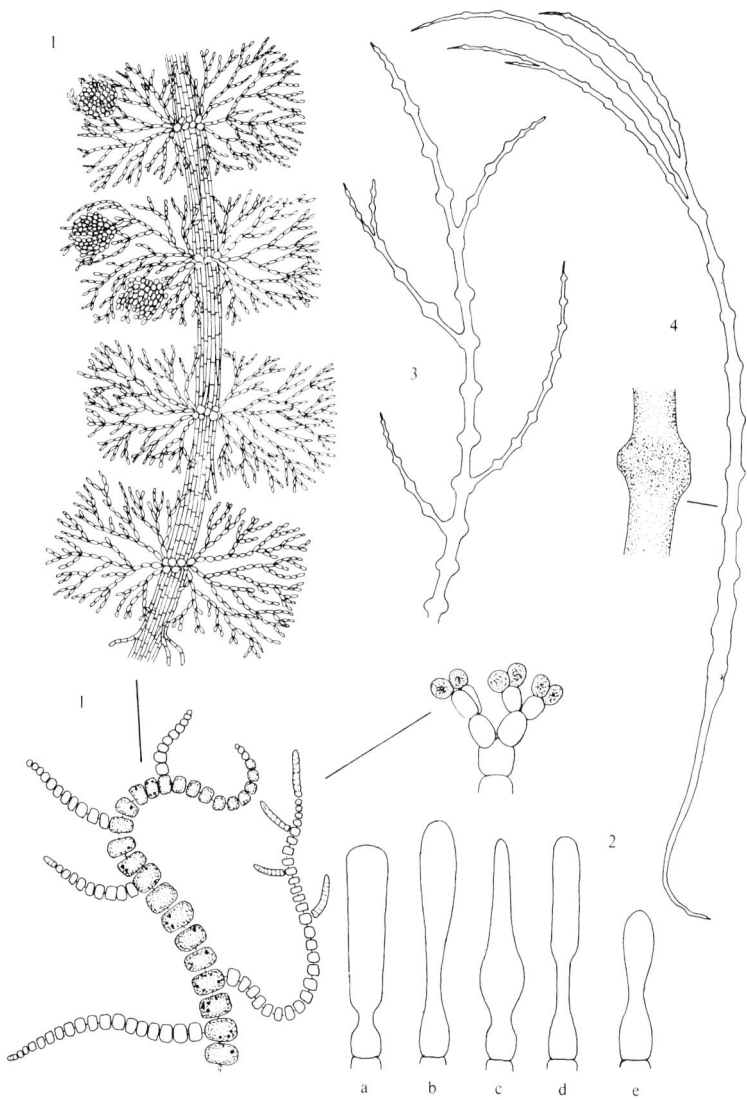

49

1b   Central axis naked, thalli often branched. Subgenus *Sacheria* 2 (Plate 10 Fig. 1b).

2a   Thalli abruptly narrowed at the base, black-brown, 6-30 cm. long, common.   *L. fluviatilis* (Plate 9 Fig. 4)

2b   Thalli gradually narrowed, olive to yellow-green.   3

3a   Dull olive-green, tufted. Male cells in prominent patches. *L. mamillosa* (Plate 9 Fig. 3)

3b   Yellow-green, not in tufts, male cells superficial. *L. fucina*

6. Nothocladus

Thalli 1-11 cm in length, 0.2-1 mm wide, dark and cartilaginous with regular annular thickenings in most species.

N. lindaueri is the largest species with thalli (2)4-11 cm long and occurs in flowing water in New Zealand. The genus seems to be confined to Australasia (Plate 10 Fig. 6, 2b).

7. Tuomeya

This is another exotic genus similar to *Nothocladus*, but it appears to be structurally distinct. *T. fluviatilis* is a small shrubby plant about 5 cm in length with well branched, blackish filaments up to 0.5 mm wide. It occurs in streams east of the Missisippi (Plate 10 Fig. 4 2c).

8. Thorea

Thalli extremely lax, consisting of numerous intertwining axial filaments giving rise to a continuous felt of long (0.3-2 mm) unbranched filaments at the surface.

PLATE 10

1 Sections through the axes of a) *Batrachospermum*, b) *Lemanea* subgenus *Sacheria*, c) *Lemanea* subgenus *Eulemanea* d) *Thorea.* x 40, 2 Sections through the axes of a) *Compsopogon*, b) *Nothocladus*, c) *Tuomeya*, d) *Bostrychia* (brackish water) genus. x 40, 3 *Lemanea torulosa* x 0.75, 4 *Tuomeya fluviatilis* x 1/2 and x 8, 5 *Compsopogon leptoclados* x 1/2, 6 *Nothocladus lindaueri*, Waitangi Falls, New Zealand x 1 and x 20, 7 *Thorea ramosissima*, Walton on Thames x 1

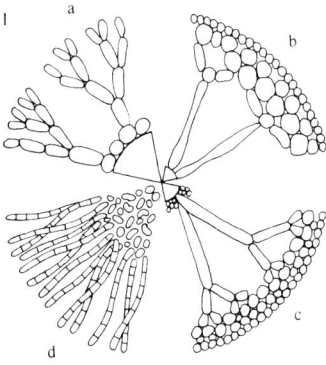

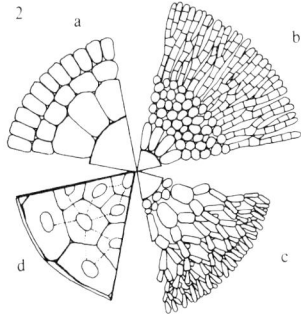

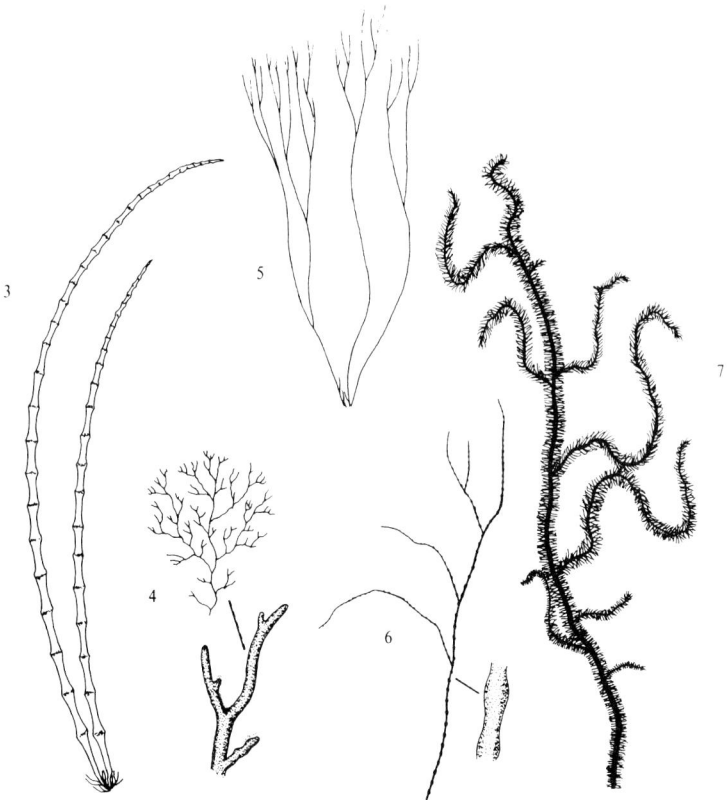

T. ramosissima grows to a length of 20-100 cm with fine, irregularly branched filaments is rare in ditches and lowland rivers in Britain. (Plate 10 Fig. 7, 1d).

9. Compsopogon

Thalli hair like and stiff in the large forms, wells branched with a trailing appearance due to the acute branch angles. Transverse sections reveal a central axis surrounded by (0)1-4 layers of rounded cortical cells. Thalli olive to brown-black (5)20-100(200) cm long, 0.05-1(5) mm wide. (Id: habit, no. of cortical layers, presence of a holdfast, 5).

The best known species is *C. chalybeus* (Plate 10 Fig. 5, 2a) with partly corticate, fine filaments, 5-20 cm long, 0.05-0.4 mm wide. This species is very rare in Britain and associated with thermal outfalls. The genus is well represented in tropical regions.

THE CHRYSOPHYCEAE

This is a large group of small and inconspicous algae with pale brown or yellowish chloroplasts. Most forms possess one or two flagella and some species are an important component of the nannoplankton, particularly in oligotrophic, softwater lakes. The nannoplankton is best examined by adding a drop of Lugol's iodine to a cylinder of lake water then examining the sediment after about three hours. Some species have their cells covered with minute silica spines which require a good lens to be seen clearly. Most Chrysophyceae produce spherical or urn-shaped silica cysts which are closed at the top by a plug (cf. the Xanthophyceae). These are probably resting stages and they are used widely in the classification of the species. The cell wall is often very thin and an outer theca or lorica is often produced. Members of this class are often found attached to other filamentous algae. Here, the flagella, when present, do not serve as swimming organs but seem to be adapted to force water currents past the cell. The free swimming flagellates often have the posterior end of their cells drawn out into a fine point. This can be useful in identification although the colour of the cells alone is usually sufficient to place them.

Key

1a   Unicellular, attached or planktonic, usually with flagella 2

1b   Colonial or filamentous, with or without flagella   9

2a   Cells flagellate and covered with minute scales or bristles, frequent in the plankton.   *Mallomonas* 10

2b   Scales absent   3

3a   Free swimming   4

3b   Epiphytes, cells with a theca   8

4a   Cells covered with a pale or brownish theca which is rounded or flattened.   5

| 4b | Theca absent | 7 |

| 5a | Cells with a flagellum passing through a pore in the theca 6 |

| 5b | Flagella absent but with a few fine strands of protoplasm passing through the pore of the theca. *Lagynion* 1 |

| 6a | Pore $> 1/3$ the diameter of the theca. *Kephyrion* 4 |

| 6b | Pore minute, hardly visible. *Chrysococcus* 2 |

| 7a | Cells with two flagella, one clearly longer than the other. *Ochromonas* 6 |

| 7b | Cells with a single flagellum. *Chromulina* 3 |

| 8a | Flagella absent, with or without fine strands of protoplasm extending out from the pore. *Lagynion* 1 |

| 8b | Cells with two flagella, one shorter than the other. Theca cylindrical and cup-shaped, opening wide, cells mostly fusiform. *Epipyxis* 8 |

| 9a | Colonial, free-swimming flagellates covered in silica scales. Cells often pear-shaped, in small, rounded colonies. *Synura* 11 |

| 9b | Scales absent. 10 |

| 10a | Colonies dendroid and planktonic, each cell within a long funnel-shaped theca and two flagella. Common. *Dinobryon* 9 |

| 10b | Not dendroid, theca absent. 11 |

| 11a | Large rounded colonies of flagellate cells, either attached or free-floating, second flagellum much reduced. *Uroglena* 7 |

| 11b | Flagella absent. 12 |

| 12a | In large, trailing gelatinous masses attached to rocks in mountain streams, with a foetid smell. Cells in irregular colonies, fusiform. *Hydrurus* 5 |

| 12b | Plants composed of short, well branched filaments. Cells barrel-shaped, with 1-several chloroplasts. Uncommon epiphyte. *Phaeothamnion* 12 |

1. Lagynion

Cells spherical with 1-2 chloroplasts and contractile vacuoles, contained within an urn- or pot-shaped theca from which fine extensions of the cell often protrude. Theca 3-6(15) µm wide, usually attached to other plants but occasionally planktonic. (Id: form and colour of theca, position of contractile vacuoles; z; 20).

L. ampullaceum (Plate 11 Fig. 8) has a small spherical theca 6-12 µm d. with a long neck-like opening and a single chloroplast.

2. Chrysococcus

Cells enclosed by a sperical theca which is smooth or ornamented with spines or ridges, 8-12(30) µm wide. Cells with two chloroplasts and contractile vacuoles. These algae are widely reported from the plankton but should not be confused with *Trachelomonas* which they resemble. (Id: theca size and structure, no. of chloroplasts, presence of stigma; 30).

C. rufescens (Plate 11 Fig. 9) has a close fitting brownish theca and two parietal chloroplasts, measuring 8-12 µm wide. It is not common.

3. Chromulina

Cells frequently asymmetrical with 1-2 chloroplasts and a single, long flagellum. The cells measure 10-20(65) µm long, ltb 1-4 and resemble *Ochromonas* in their shape. The species are found in small numbers, usually among vegetation. (Id: cell shape, number of chloroplasts, presence of stigma, structure of cysts; 100).

C. nebulosa (Plate 11 Fig. 10) is a species with a smooth, asymmetric cell wall and a single parietal chloroplast without a pyrenoid. There is no stigma and the cells measure 8-16 µm long. The cysts, if formed, are spherical, about 10 µm wide with fine longitudinal striations.

4. Kephyrion

Cells enclosed within an urn-shaped theca, pale or brownish in colour and (2)5-8(15) µm wide. The theca is smooth or has a series of annular or helical bands with an anterior pore through which the flagellum passes. These minute flagellates occur in the plankton but are easily overlooked. (Id: theca structure; I; 30).

K. starmachii (Plate 11 Fig. 11) is a small species with a pot shaped lorica and three annular thickenings, diameter 4-5 µm. Not yet recorded from Britain.

5. Hydrurus

Cells spheroidal to fusiform with one chloroplast and sometimes several contractile vacuoles, 8-15 µm long, ltb 1.5-3. Loosely packed in large mucilaginous masses occurring as brown feathery strands in upland streams.

H. foetidus is the only species and is rare in northern Britain (Plate 12 Fig. 1). A similar alga, *Chrysonebula holmesii* forms thick layers of pale jelly in calcareous streams of the north. The cells are spherical, 3-6 µm wide with a single chloroplast and no pyrenoid (Plate 12 Fig. 6).

6. Ochromonas

Cells fusiform, somewhat asymetric with two subapical flagella, 1-several chloroplasts and frequently with two contractile vacuoles and a stigma. The cells measure 5-20(30) µm long, ltb 1-2(3) and are normally found in small numbers among vegetation. Some species grow well in the laboratory where they have been used widely as experimental organisms. (Id: cell shape, presence of a pyrenoid, number of chloroplasts, form of cysts, flagellum length; 50).

O. pyriformis is a distinctive species with two small chloroplasts and a long, narrow 'tail'. Cells 11-18 µm long, ltb 3-4 (Plate 11 Fig. 13).

PLATE 11

1 *Tribonema viride*, Malham Tarn x 900, 2 *T. minus* Malham Tarn x 900, 3 *Characiopsis polychloris*, Cuckmere, Sussex x 1275, 4 *C. turgida* x 450, 5 *Mischococcus confervicola*, Newhaven, Sussex x 750, 6 *Ophiocytium arbuscula* x 375, 7 *O. parvulum*, Southborough, Kent x 750, 8 *Lagynion ampullaceum* x 1725, 9 *Chrysococcus rufescens* x 1725, 10 *Chromulina nebulosa*, College Valley, Northumbria x 2250, 11 *Kephyrion starmachii* x 3500, 12 *Epipyxis utriculus* cell and theca x 750, 13 *Ochromonas pyriformis* x 2250

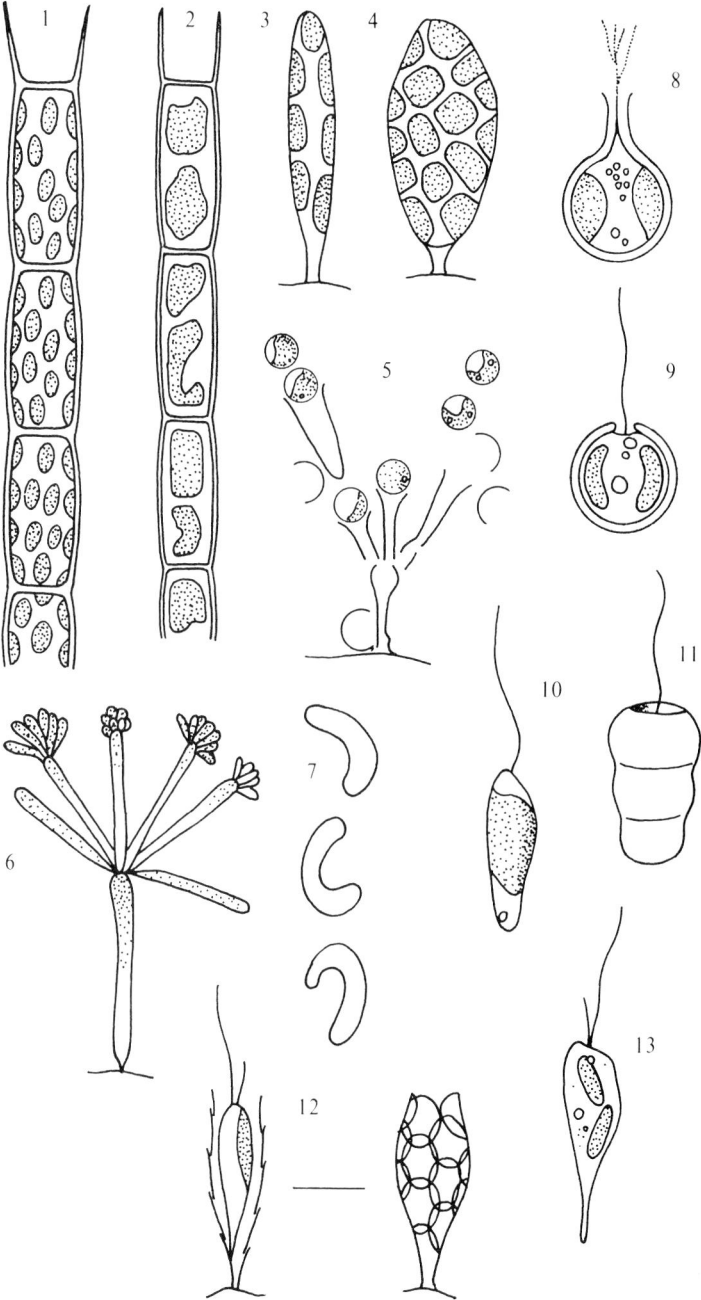

7. Uroglena

Cells in large globular colonies up to 1 mm in diameter, attached to rocks or planktonic. The cells are pear-shaped or fusiform, (5)10-18(22) µm long, ltb 1.5-3, with 1-2 chloroplasts. The cells sometimes appear to be connected to their neighbours by delicate mucilagonous threads. Species of *Uroglena* occasionally occur in the plankton in large numbers but little is known of their ecology. (Id: form and structure of cysts, presence of mucilage strands; 10).

U. americana (Plate 12 Fig. 2) has smooth spherical cysts about 10 µm wide and cells 5-12 µm long whilst *U. volvox* has cysts with a broad cylindrical collar surrounding the plug. The cells are 12-20 µm long, often attached by fine strands of mucilage (Plate 12 Fig. 3).

8. Epipyxis

Cells surrounded by a cylindrical theca composed of minute imbricating scales with a broad apical opening. The cells are narrowly pear-shaped and attached to the theca by a fine strand and measure 8-30(50) µm long, ltb 4-6. The species are usually solitary but sometimes occur in small groups and seem to be most abundant in oligotrophic waters. (Id: theca shape and structure; 30).

E. utriculus (Plate 11 Fig. 12) has cells 22-50 µm long, ltb 4-5 with a single parietal chloroplast. The theca is composed of about a dozen oval, overlapping scales 5-8 µm in length. The scales are organic in nature and not silicified.

9. Dinobryon

Cells enclosed in thimble-like thecae joined together to form

---

PLATE 12

1 *Hydrurus foetidus*, Dordogne, France x 1.5 and x 225, 2 *Uroglena americana*, colony x 700, cyst x 1300, 3 *U. volvox* cyst x 825, 4 *Synura uvella*, Speldhurst, Kent. Colony x 450, cell x 1200 and scale x 4500, 5 *S. sphagnicola* scale x 4500, 6 *Chrysonebula holmesii*, Arncliffe, Yorks, part of colony with calcite crystals, x 1000, 7 *Dinobryon divergens*, Speldhurst, Kent. Colony x 375, cell x 1125 and cyst x 1125, 8 *Mallomonas acaroides*, Windermere, Cumbria, cell x 825 and scale x 3750.

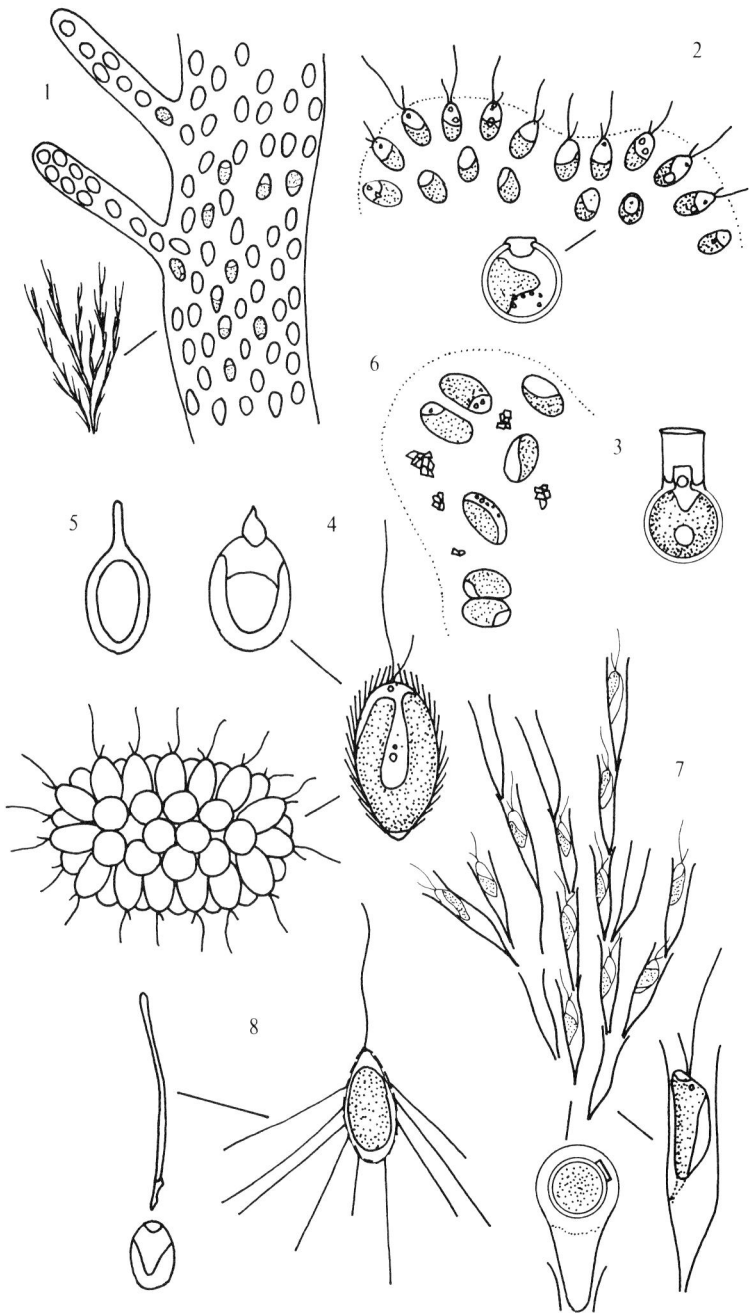

dendroid colonies. *Dinobryon* is the best known member of this class in freshwater and it is widely distributed and frequently abundant. The colonies contain up to 50 cells with the thecae measuring (10)15-60(80) µm long, ltb (3)4-6(9). The cells are fusiform and the flagella protrude from the wide thecal opening. Each cell has a single parietal chloroplast and usually a prominent apical stigma. Laboratory studies suggest that *Dinobryon* is highly sensitive to certain dissolved substances, particularly phosphate and perhaps calcium and magnesium. The species often develop in oligotrophic lakes but they sometimes also occur in eutrophic waters as well. Further work is needed to demonstrate the sensitivity of the species conclusively. (Id: colony shape, theca size and shape; z(4); 20).

D. divergens is the best known species with diverging colonies and thecae 30-65 µm long, ltb 3-5 (Plate 12 Fig. 7). Spherical cysts are frequently produced, 10-14 µm wide. The difference between many of the species is very small.

10. Mallomonas

Cells elongate-cylindrical, covered in numerous, overlapping silica spines, with an apical flagellum and 1-2 parietal chloroplasts. The cells measure (8)12-30(100) µm long, ltb 1.5-4(6). The spines are frequently detached from the cells and are normally shaped like a shield attached to a long, narrow needle. The species occur in small numbers in the lake plankton. (Id: spine position on the cells, spine shape, size and shape of cysts; I; 60).

M. acaroides is frequently recorded in Britain. The cells measure 18-45 µm long, ltb 1.5-2 and are egg-shaped. Spines shield-like, 6-7 x 4 µm with V-shaped markings (Plate 12 Fig. 8). The cysts are spherical and minutely granular with a small plug.

PLATE 13

1 *Phaeothamnion confervicolum* x 400, 2 *Cryptomonas ovata* Wicken Fen, Cambs. x 1750, 3 *C. tetrapyrenoidosa* x 1350, 4 *Chroomonas nordstedtii* x 2500, 5 *Trachelomonas volvocina*, E. Grinstead, Sussex x 1100, 6 *T. hispida*, E. Grinstead, Sussex, x 1100, 7 *Lepocinclis fusiformis*, Newhaven, Sussex x 1500, 8 *Phacus suecicus*, Eridge, Sussex x 1100, 9 *Euglena mutabilis*, Tunbridge Wells, Kent, x 1000, 10 *E. viridis* x 900, 11 *E. spirogyra* x 250, 12 *E. gracilis* x 550, 13 *E. tripteris* x 900

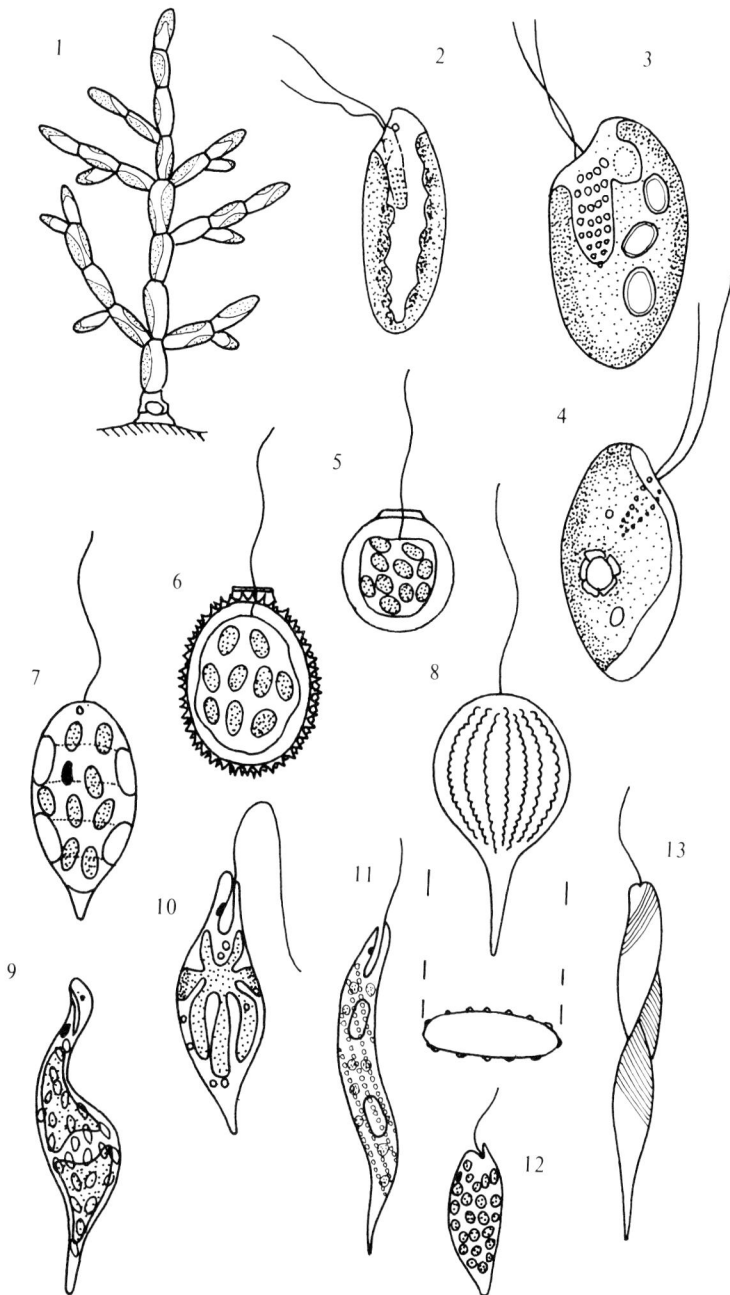

61

11. Synura

Rapidly swimming, golden-yellow colonies of *Synura* are often seen in the plankton of small lakes and ponds. Like *Mallomonas*, the cells are covered in silica scales but they possess two apical flagella, not one. The cells of the colonies number about 10-40 and are close-packed and generally pear-shaped, and 8-20(30) µm long, ltb (1.4)2-3(5). The scales are normally smaller than those of *Mallomonas*. (Id: shape of the chloroplasts, presence of a stigma, shape of the cysts, structure of scales; I; 10).

S. uvella (Plate 12 Fig. 4) has pear-shaped cells 20-40 µm long, ltb 2.5-3.5 with two parietal chloroplasts. The colonies measure 80-400 µm long and the scales are shield-like with a tooth-like projection at one end. They measure 4-6 µm in length. *S. sphagnicola* is similar but the scales are elliptical and the tooth is longer giving the appearance of a tennis raquet (Plate 12 Fig. 5).

12. Phaeothamnion

Plants consist of small, well-branched filaments composed of barrel-shaped cells 6-12 µm wide containing 1-several small discoid chloroplasts. They are found attached to aquatic vegetation, usually in small lakes or ditches. The species are rarely seen but widely distributed and perhaps associated with *Lemna, Cladophora* and *Vaucheria*. (Id: chloroplast number, cell shape, habit; 5).

P. confervicolum (Plate 13 Fig. 1) has regularly branched filaments and grows up to 250 µm long. The cells are 20-30 µm long, ltb 2-4 with a single chloroplast per cell.

XANTHOPHYCEAE

This class is superficially similar to the Green Algae but is distinguished by the lack of a starch reaction, the two unequal flagella and the occurrence of spherical silica cysts with lid-like openings. Other useful but not always reliable features are the presence of several discoidal chloroplasts per cell and the blue-green reaction of the chloroplasts on the application of dilute hydrochloric acid. In the absence of flagellate stages, these differences are quite small when it is realized that a positive starch test for Green Algae is not always given. There remain at the time of writing a number of algae whose classification remains undecided.

The species of Xanthophyceae are comparatively small in number but they contain several abundant filamentous species. *Vaucheria* is easy to identify because of the lack of cross walls, whilst *Tribonema*, which is equally common has a peculiar and distinctive wall structure. Many forms are associated with nutrient-rich waters.

Key

1a Plants spherical, up to 3 mm diameter, rooted by rhizoids into soft mud and consisting of a single large cell.
Botrydium 4

1b Plants filamentous or colonial, if single celled, then small. 2

2a Filamentous   3

2b Single celled or colonial   4

3a Plants consisting of long tubular filaments without cross walls.   *Vaucheria* 6

3b Filaments with cross walls and the end-cells or fragments H-shaped.   *Tribonema* 5

4a Unicellular epiphytes, normally attached by a small pad of mucilage.   *Characiopsis* 1

| 4b | Algae colonial or if unicellular then planktonic. | 5 |

| 5a | Cells spherical forming dendroid colonies attached by strands of mucilage. | *Mischococcus* 2 |

| 5b | Cells cylindrical or sausage-shaped, with rounded or finely pointed ends, unicellular or colonial and umbrella-shaped. | *Ophiocytium* 3 |

1. Characiopsis

Cells with rounded or pointed apices and attached to the substratum by a narrow mucilage stalk. The cells contain several discoid chloroplasts without pyrenoids and measure (5)10-35(90) µm long, ltb (1)3-6(15) and they are common epiphytes, although some 'species' might be germinating zoospores of other Xanthophyceae. A similar genus, *Characidiopsis* is distinguished by the presence of contractile vacuoles and a stigma. (Id: shape of cell, length of stalk, number of chloroplasts, habitat; z; 20).

C. turgida is a species with plump cells as the name suggests (Plate 11 Fig. 4) it is 30-50 µm long, 1.2-1.5 with 10 or more chloroplasts, whilst *C. polychloris* (Plate 11 Fig. 3) has more slender cells, 20-35 µm long, ltb 4-7 with 8 or more chloroplasts. Both occur in Britain.

2. Mischococcus

Cells spherical, rarely spheroidal with one to several pale discoid chloroplasts, (2)4-8(10) µm wide. The cells are often paired into dendroid groups held together by mucilage which forms a gelatinous stalk. Old cell wall remains are often seen close by. These minute algae are locally frequent on aquatic vegetation, often in ditches near the sea. (Id: cell shape, colony structure; z; 4).

PLATE 14

1 *Vaucheria sessilis*, Sheppey, Kent, with oosphere and dehisced antheridium. x 260, 2 *V. geminata* x 110, 3 *V. glomerata* with 3 dehisced antheridia x 45, 4 *V. pachyderma* x 75, 5 *V. ornithocephala* x 80, 6 *V. debaryana* x 75, 7 *Botrydium granulatum*, Tonbridge, Kent, x 10, 8 *Phycopeltis arundinacea*, on Ivy, Slaney Estuary, Co. Wexford, x 90, 9 *Trentepohlia aurea*, Arncliffe, Yorks, habit x 40, branch and sporangium x 400.

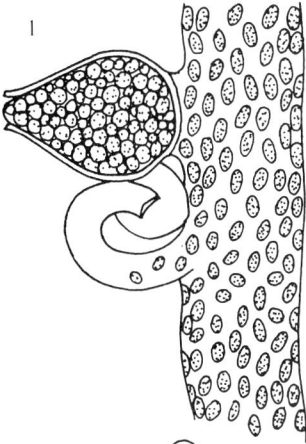
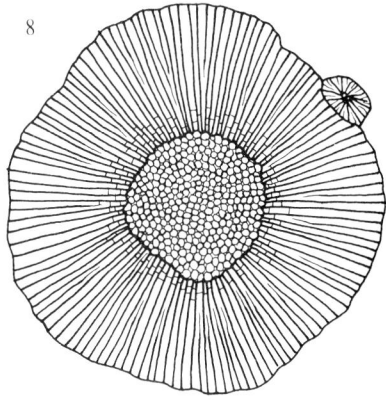
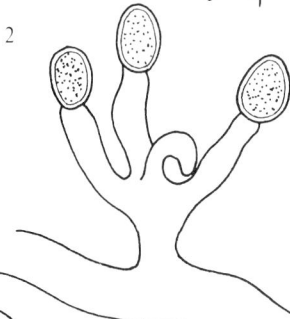
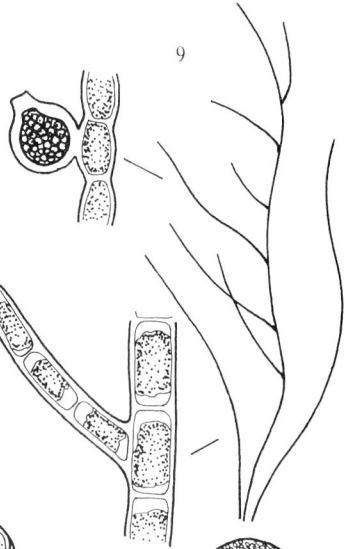
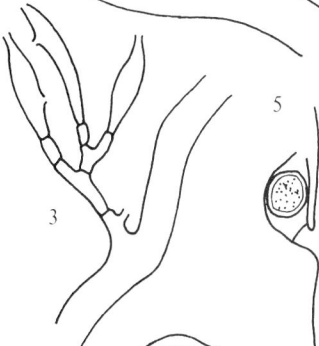
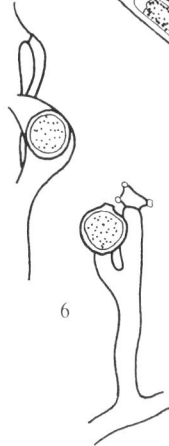
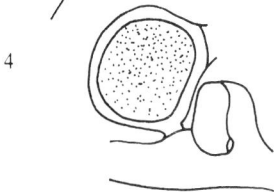
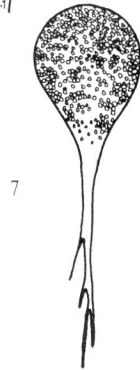

65

M. confervicola is the commonest form with spherical cells 5-8 μm wide and 2-8 chloroplasts. It forms small dendroid colonies (Plate 11 Fig. 5).

3. Ophiocytium

Cells cylindrical and often much twisted with rounded or spine-like ends, (5)10-80(600) μm long, ltb 4-15(25). The cells contain several discoid chloroplasts without pyrenoids and they occur as epiphytes, or, more often, free-floating among water plants. None of the species is particularly common although they are sometimes numerous among the bladderwort (*Utricularia*). (Id: presence of end-spines, colony structure, cell shape; z; 10).

O. arbuscula is a colonial species consisting of an upright basal cell which sprouts a tuft of more cells at the apex. These cells may give rise to a further tuft resulting in a tree-like appearance (Plate 11 Fig. 6). The cells measure 60-120 μm long, ltb 10-15. *O. parvulum* is a commoner, unicellular form with strongly twisted sausage-shaped cells 4-6 μm wide and devoid of spines (Plate 11 Fig. 7).

4. Botrydium

B. granulatum (Plate 14 Fig. 7) is the best known species with large pear-shaped cells, 0.5-3 mm diameter, partly buried in mud. The cells contain numerous discoid chloroplasts with pyrenoids and branched, colourless rhizoids anchor the plant to the sediment (I, z, Aut; 5). *Botrydium* can develop in extensive patches on recently exposed mud but it is not often seen in Britain. *Protosiphon* is a similar alga belonging to the Chlorophyceae and has a reticulate chloroplast and many pyrenoids, but appears to be much less common than *Botrydium*.

5. Tribonema

Filaments unbranched, (2)4-12(25) μm wide, cell ltb (1.5)2-6(12) with 1-many rounded chloroplasts per cell, usually without pyrenoids. The filaments, in common with *Microspora* (Chlorophyceae) have a wall structure with a median suture which causes breakages to occur in the middle, forming H-shaped ends and fragments. The cells are often barrel-shaped and the species are found in abundance in shallow water. (Id: number of chloroplasts, presence of pyrenoids; z; 15).

The commonest species appears to be *T. viride* which can block filters in water treatment plants. The cells are 8-20 µm wide, ltb 2.5-8 and contain numerous chloroplasts (Plate 11 Fig. 1). *T. minus* (Plate 11 Fig. 2) is a much smaller plant with cells 3-6 µm wide, ltb 2-3 and with just 1-3 chloroplasts per cell.

6. Vaucheria

Plants composed of long, tubular filaments, (15)30-250(500) µm wide, frequently found in dense mats or felts in shallow water or on mud. The species often have a rough feel like *Cladophora*, in contrast with most other algae. The chloroplasts are small and numerous, without pyrenoids. These algae have an unusual method of sexual reproduction and deserve much further study. They are all oogamous and most individuals are bisexual whilst a few have the sexes on different plants. The sex organs begin as small outgrowths along the filament. The female outgrowths are simple bulges or short, branched structures bearing one or two dark green, spherical oogonia which often turn red after fertilization. The male outgrowths are narrower and develop close to the females. They are usually cylindrical and twisted and seated in some species on a support cell cut off by a transverse wall. This outgrowth produces numerous multiflagellate sperms which enter the female cell via a small pore or slit. The fertilized egg often germinates directly on the parent to form a new plant. About 60 freshwater species have been described on the basis of the structure of the reproductive organs. The species are often found fertile and they are usually seen in shallow, hard water particularly during the winter. The genus is subdivided into several sections and some common representative species are described below.

1a   Antheridia seated upon a support cell, elongate-cylindrical with terminal and lateral pores.
        (Piloboloideae). *V. glomerata* (Plate 14 Fig. 3) has not yet been recorded from Britain. The antheridia have 1-2 lateral pores and there may be several developing from a single

support cell. The oogonia are found on short side branches and the oospores measure 100-160 µm long. Filaments 20-50 µm wide.

1b   Support cell absent, antheridia variously shaped, often twisted, with 1-several pores or slits.   2

2a   Antheridia cylindrical and usually lying against the filament, opening by a slit (Tubuligerae).
     V. ornithocephala (Plate 14 Fig. 5) has filaments 20-50(75) µm wide. The oospores are 80-150 µm long and completely fill the oogonium.

2b   Antheridia arising vertically from the filament, sometimes developing on a non-septate side branch.   3

3a   Antheridia not twisted   4

3b   Antheridia strongly twisted in one plane (Corniculatae). This is the largest group with several common species.
     V. sessilis (Plate 14 Fig. 1) belongs to a subgroup with erect, sessile oogonia and a single pore often directed a little to one side and filaments 30-120 µm wide.
     The remaining subgroup is characterised by the sex organs appearing on long side branches. *V. terrestris* has filaments 40-100 µm wide. The distinguishing character is the form of the side branch where the oogonium develops above the antheridium and measures 80-210 µm in length. *V. geminata* (Plate 14 Fig. 2) has side branches supporting (1)2-6 oogonia. The oogonia measure 70-110 µm long and possess a 3-layered wall often covered with brown spots. Filaments 40-80 µm wide. *V. hamata* is similar but there are normally 1-2 oogonia per branch. The oospores are grey in colour and 70-90 µm long, filaments 40-80 µm wide.

4a   Antheridia cylindrical or spherical with an apical pore (Globiferae).
     V. pachyderma (Plate 14 Fig. 4) has filaments 40-120 µm wide. The oogonia are depressed sideways, 70-220 µm long with a wide wall 6-7 µm thick.

4b Antheridia pear-shaped or angular, with several lateral pores (Anomalae).

An example is *V. debaryana* (Plate 14 Fig. 6) with filaments 20-60 μm wide and oogonia in groups of 1-3, oospores 50-80 μm long and short erect antheridia with 1-5 pores.

THE DINOPHYCEAE

This is a class of important marine flagellates with a small number of freshwater representatives. They are a well defined group with many unusual features. In common with the Cryptophyceae, the chloroplasts may possess a variety of pigments although the cells are usually brownish in colour. Some forms also possess trichocysts (p.79). Most dinoflagellates are rounded but somewhat flattened cells with two prominent furrows. One of these encircles the cell and may be faintly helical, whilst the other breaks off from this transverse furrow and ends near one apex. The two furrows contain the flagella which differ greatly in form and function. The transverse flagellum has a large number of helical turns and a peculiar 'winged' structure when viewed in the electron microscope. The other, longitudinal flagellum is similar to the flagella of other algal classes and propels the cell from behind, whilst the transverse flagellum imparts a rotating force which improves swimming speed, and often propels the cell forwards on its own. The transverse furrow divides the cell into two regions. The anterior end, or *epicone* and the posterior end or *hypocone* which contains the longitudinal furrow.

The common freshwater forms have been divided into two orders. This division separates species with a distinctly ornamented cell wall composed of angular plates and those with a smooth, naked periplast. However, studies with the electron microscope suggest that no clear-cut division of this kind is possible since the two types grade into each other.

The cells possess 1-many disc-like chloroplasts and an orange stigma is often evident. Thick-walled cysts are frequently produced, but isogamous sexual reproduction is rarely observed. Starch is a common storage product. Details of the wall structure can sometimes be revealed by the addition of a dilute bleach solution to the sample. The cells of some genera, particularly *Gymnodinium* and *Ceratium* often cease movement soon after collection and they can rapidly disintegrate when viewed under the microscope.

Key

1a	Cells naked, with a smooth periplast.	2
1b	Cells covered with angular plates.	3
2a	Epicone and hypocone about the same length.	*Gymnodinium* 1
2b	Epicone and hypocone very different in length	*Amphidinium* (rare) 2
3a	Cells rounded, slightly laterally flattened.	*Peridinium* 3
3b	Cells strongly flattened, with 3-4 horn-like spines.	*Ceratium* 4

1. Gymnodinium

Cells a little flattened and spheroidal measuring (8)15-45(130) µm long, ltb 1-1.5(2.3) with an approximately median transverse furrow, these helmet-shaped organisms are sometimes found in abundance in the plankton but are most often seen in small numbers in ponds and ditches. (Id: chloroplast colour, shape of cell; 50). The species are difficult to identify since the shape seems to be quite variable.

G. inversum (Plate 15 Fig. 1) is a medium-sized species measuring 25-35 µm in length, ltb 0.9-1.7. It is only slightly flattened laterally and has rounded apices. The chloroplasts are yellow-brown in colour. *G. aeruginosum* (Plate 15 Fig. 2) is one of the few species with bright blue-green chloroplasts. The cells measure 30-35 µm long, ltb 1.5-1.7 and are considerably flattened. It is not common.

2. Amphidinium

This is a genus similar to the preceeding but the hypocone is 3 or more times longer than the epicone. The cells measure (10)12-35(50) µm in length, ltb (0.8)1-2.3(2.7) but they are far less frequently seen than *Gymnodinium*. (Id: chloroplast number and colour, cell shape; 20).

A. operculatum is a large species with an extremely small, almost lid-like epicone whilst the hypocone is elongate spheroidal in shape. The chloroplasts are brown and the cells measure 35-50 µm in length, ltb 1.8-2.2 (Plate 15 Fig. 6).

3. Peridinium

Cells covered in angular plates giving a polygonal appearance and sometimes with small patches of sharp spines. Transverse furrow often a little offset from the median position so that one cone is often slightly larger than the other. There are usually 7 plates on the hypocone and about 13 plates on the epicone. Cells measure (15)25-50(100) µm long, ltb 0.9-1.5. Identification is a tedious process since it is based upon the number and arrangement of the plates although some biologists believe that this is unlikely to be an inherited character. One group can be identified by the presence of a small apical pore on the epitheca. About 30 freshwater forms have been described and 3 common British species are described below.

Key to species illustrated

1a A small pore present on the apex of the epicone. Cells rounded, 15-30 µm long, ltb 1-1.3. Epicone plates number about 14, sometimes with a few short spines on the hypocone.   *P. inconspicuum* (Plate 15 Fig. 5)

1b Pore absent, epicone plates 14, cells rounded but may be laterally flattened.   2

2a Transverse furrow with a pronounced raised edge, slightly asymmetric. Cells 45-70 µm long, ltb 0.9-1.3. *P. willei*
(Plate 15 Fig. 4)

PLATE 15

1 *Gymnodinium inversum*, College Valley, Northumbria x 1200, 2 *G. aeruginosum* x 1300, 3 *Peridinium cinctum* Wicken Fen, Cambs. x 600, 4 *P. willei* x 600, 5 *P. inconspicuum*, side view and epicone with pore, x 1000, 6 *Amphidinium operculatum* x 800, 7 *Ceratium hirundinella*, Windermere x 300 and cyst x 110, 8 *Vacuolaria virescens*, Malham Fen, Yorks x 1200

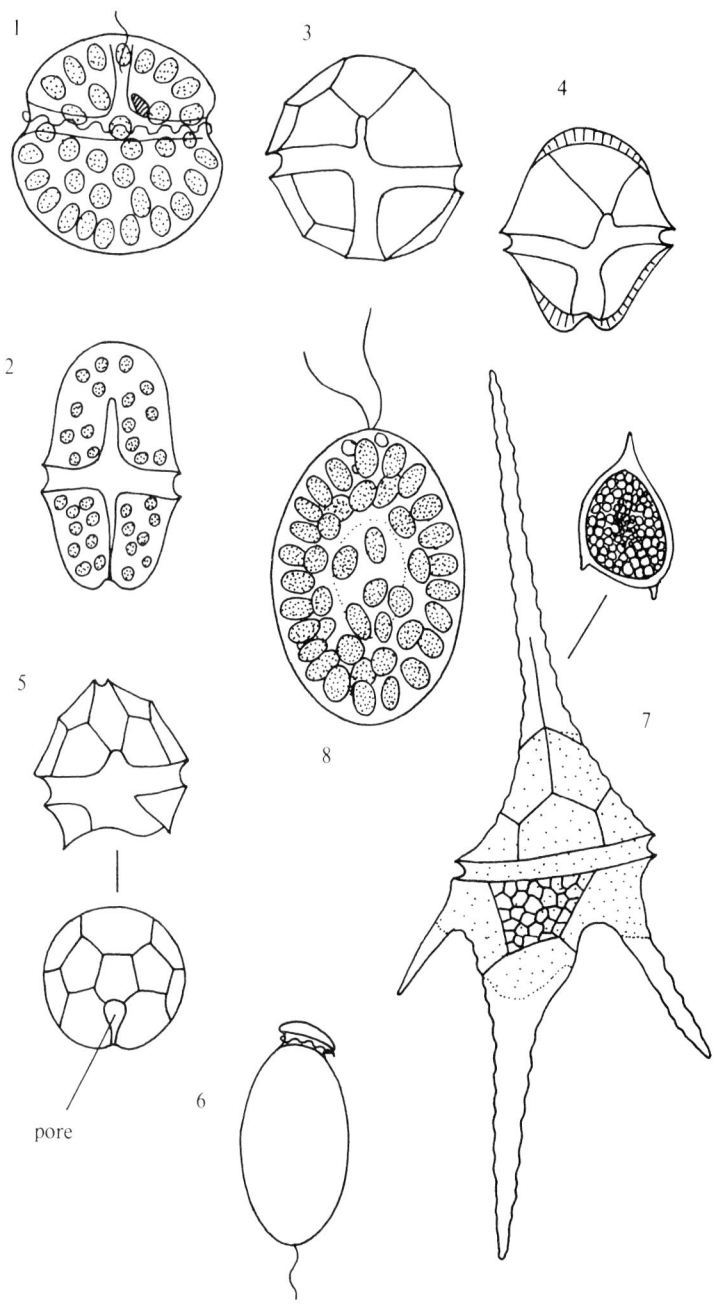

1

3

4

2

8

7

5

pore

6

73

2b   Transverse furrow broad, edges not raised prominently, epicone and hypocone equal in size. Cells 35-55 µm long, ltb 0.9-1.1.   *P. cinctum* (Plate 15 Fig. 3).

4. Ceratium

Cells large, with one anterior and 2-3 posterior processes. The transverse furrow is narrow and in an approximately median position, but the longitudinal furrow is much less evident. The cells are covered by a series of plates which have a fine and complex areolation and the cells are strongly flattened and may be occasionally somewhat curved when viewed edge-on. The chloroplasts are yellow-brown, small and discoidal. Cysts are commonly produced towards the end of the year with thick walls and spines at the angles.

There is one well known and quite variable species, *C. hirundinella* with cells (40)60-300(400) µm long, ltb (1.5)3-4(8). (Plate 15 Fig. 7). This is a common planktonic alga which can produce a dark brown colouration in eutrophic lakes. The vertical migrations of this species in response to light and nutrients has been well documented.

EUGLENOPHYCEAE

This is a large class of mostly unicellular flagellates of considerable interest. Some members can lose their chloroplasts and find alternative sources of energy, whilst others rely entirely upon other sources but these types are not considered here. The basic features of the class are a 'naked' protoplast, such as that seen in *Amoeba*, supported internally by bands of protein, termed the *pellicle*. This structure is sometimes seen as a series of helical striations under the microscope (Plate 13 Fig. 13). There are normally two flagella which arise from a pronounced depression or pit near the cell apex. This pit is termed the *reservoir* and only one flagellum is normally seen emerging from it. The other is extremely short and not often seen and it appears to act as a light receptor – nature's answer to the optical fibre – where light seems to be guided to control the direction in which the cell moves. A few species have an outer covering, or theca which is often dark brown in colour.

Euglena sometimes exhibits 'metabolic' behaviour. This is a pronounced and rapid change of shape brought about by internal movements and an elastic pellicle but not all species show it. The Euglenoids possess several discoid chloroplasts, sometimes with pyrenoids but the range of shape is wide. The storage product is paramylum. This substance occurs as small refractive granules whose shape is used in classification. Unlike the Chlorophyceae, the Euglenoids have a conspicuous red stigma near the flagellum base which is not associated with chloroplast.

The species are found in a wide variety of habitats but usually where there are accumulations of organic matter. Pools among manure are often covered with a green or reddish scum composed of several species growing together. *Trachelomonas* is probably the most widespread form and occurs in all kinds of habitats and not necessarily those which have been organically polluted. Sexual reproduction is unknown in the group but resting cysts are frequently encountered.

Key

1a   Cells contained within a spherical or spheroidal, brownish theca with a pore sometimes surrounded by a small collar.

Trachelomonas 1

1b   Cells without a theca   2

2a   Cells flattened and often with a striated pellicle. *Phacus* 2

2b   Cells roughly circular in section, often fusiform.   3

3a   Paramyum reserves composed of one or two annular rings close to the pellicle.   *Lepocinclis* 3

3b   Paramylum rod-shaped or rounded in small granules. Cells often metabolic.   *Euglena* 4

1. Trachelomonas

Cell enclosed within a rounded theca from pale to dark brown in colour, with or without spines and ridges. The flagellum passes out through the collared pore and connects with the cell within. Chloroplasts several, discoidal with a small stigma usually visible. The thecae measure (4)10-30(50) µm in length, ltb 1-2.5 and the species are abundant in most waters, but particularly in small ponds. There appear to be a few extremely polymorphic species whose theca ornamentation and size depends upon the amount of reduced iron and manganese in the water, but many have been described. (Id: theca shape and ornamentation, development of collar; 60).

T. volvocina (Plate 13 Fig. 5) has a smooth, spherical theca without a collar and (5)8-20 µm wide. It is very common. *T. hispida* is less frequent and has a spheroidal theca 20-40 µm long, ltb 1.1-1.4 covered with numerous short spines (Plate 13 Fig. 6). These algae can be confused with *Coccomonas* and *Chrysococcus*.

2. Phacus

Cells strongly flattened with a broad anterior end and finely tapered posterior end, measuring (15)25-50(180) µm long, ltb

(1.1)1.3-1.8(4) and normally with a finely striated pellicle. Chloroplasts small, numerous and discoidal. The species occur with other Euglenoids associated with decaying vegetation or in farm ponds. (Id: cell shape, pellicle structure, shape of paramylum; 100).

P. suecicus has a roughly longitudinal series of warty striations and a finely tapered posterior end, measuring 34-36 µm long, ltb 1.2-1.6 (Plate 13 Fig. 8). The paramylum occurs as four large plates.

3. Lepocinclis

Cells fusiform and circular in section, with a helically striated periplast and with a broad anterior end and tapered posterior end. Chloroplasts numerous and discoidal. The paramylum bodies consist of two large annular rings, one above the other and the cells measure (20)25-40(60) µm long, ltb 1.5-2.5. The species occur in small numbers in organically enriched sites. (Id: number of paramylum rings, structure of periplast, length of tail; 40).

L. fusiformis has broadly fusiform cells, 20-40 µm long, ltb 1.6-2, a helically striated periplast and two paramylum rings (Plate 13 Fig. 7).

4. Euglena

Cells usually fusiform with an apical reservoir and prominent stigma and contractile vacuoles nearby. Chloroplasts 1-many, discoidal, stellate, plate-like or in long helical bands, with or without pyrenoids. Paramylum granules generally numerous, composed of discs or rods distributed throughout the cell. The helical pellicle is usually visible with a good lens. Cells measure (12)30-150(500) µm long, ltb (1)4-12, often swimming rapidly, with the flagellum passing around the cell like a maypole rope. The species are often abundant in small pools, forming green or reddish blooms. (Id: cell shape, shape and number of chloroplasts, shape of paramylum bodies, degree of metabolic activity, presence of muciferous bodies which stain in Neutral Red dye; 100).

Key to illustrated common species

1a Flagella not evident but cells highly metabolic, creeping in an amoeboid fashion with two large chloroplasts and numerous paramylum granules. Cells 15-50 µm long, ltb 1-6. Common in acidic waters.   *E. mutabilis* (Plate 13 Fig. 9)

1b Flagella present   2

2a Cell with a single lobed chloroplast, metabolic, with a finely striated periplast, 50-70 µm long, ltb 3-5.
                                          E. viridis (Plate 13 Fig. 10).

2   Cells with many chloroplasts   2b

3a Periplast with helical rows of pearly granules. Cells slightly metabolic, 80-150 µm long, ltb 6-10.   *E. spirogyra* (Plate 13 Fig. 11).

3b Pearly granules absent   4

4a Cells non-metabolic, periplast finely striated and with 2 large rods of paramylum, 60-80 µm long, ltb 6-8.
                                          E. tripteris (Plate 13 Fig. 13)

4b Cells metabolic and stout, 35-50 µm long, ltb 2-5.
                                          E. gracilis (Plate 13 Fig. 12)

THE CRYPTOPHYCEAE

This is a small class of bean-shaped flagellates with two slightly unequal flagella inserted obliquely near the apex. A small depression or furrow is developed below which is lined with minute pore-like structures known as trichocysts. These bodies are also found in the Chloromonadophyceae but they are only clearly seen in the electron microscope. The common forms possess either one or two large parietal chloroplasts which are olive or occasionally reddish or blue in colour. The strong starch reaction and the shape of the cells is a good guide to the class. The species are common in all kinds of habitats and important members of the nannoplankton. However, due to their small size, they are not easy to identify and are often overlooked.

Key to genera

1a   Flagellar depression with many rows of trichocysts visible (good microscope lens). Chloroplasts with 0-4 pyrenoids.
Cryptomonas

1b   Flagellar depression or furrow with at most two rows of trichocysts. Chloroplasts 1-2, usually with a pyrenoid.
Chroomonas

1. Cryptomonas

Cells elliptical to bean-shaped, usually slightly laterally flattened, (10)15-30(80) µm long, ltb (1.5)2-3(3.8) with 1-2 chloroplasts, sometimes with pyrenoids. The cells are usually olive or brownish in colour but the flagellar depression can only be seen clearly in the larger species. These algae occur in small numbers in most water bodies but they have been little studied. (Id: chloroplast colour and number, number of pyrenoids, size of the depression; 40).

C. ovata has been widely reported in Britain. This species has broadly rounded cells measuring (14)20-40(80) µm long, ltb 1.9-2.3 with olive or brown chloroplasts without pyrenoids. (Plate 13 Fig. 2). *C. tetrapyrenoidosa*, as its name suggests, has four large pyrenoids (Plate 13 Fig. 3) and two olive-brown chloroplasts. The cells are 20-50 µm long, ltb 3-5. Species with pyrenoids appear to be much less common than those without them in Britain.

2. *Chroomonas*

The cells of *Chroomonas* are generally considerably smaller than those of *Cryptomonas*, measuring (7)10-18(23) µm long, ltb (1.5)2-3(3.5). There are 1-2 chloroplasts, blue or reddish in colour usually with a pyrenoid. The flagellar furrow is often difficult to discern. The cells are sometime comma-shaped and may be abundant in the nannoplankton. (Id: as *Cryptomonas*; 40).

C. nordstedtii (Plate 13 Fig. 4) has elongate-spheroidal cells 8-20 µm in length, ltb 1.6-3 and a short faint row of trichocysts. There is a single, lateral, bluish-green chloroplast and a pyrenoid. It is widely distributed.

CHLOROMONADOPHYCEAE

This is a small and little known class of flagellates. Their distinguishing features are a soft, elastic periplast which frequently allows metabolic movements, as in *Euglena*, an apical depression from which two long flagella arise and numerous discoid chloroplasts. The cells are usually found swimming with one flagellum directed forwards and the other backwards and the storage products are oils. Only two members of this class are at all frequent. They are found among mud and aquatic vegetation.

Vacuolaria virescens (Plate 15 Fig. 8) has large, elongate-spheroidal cells which are slightly flattened in section, measuring 50-160 µm long, ltb 2-3. The apical depression is not well developed but there are usually several large contractile vacuoles near the base of the flagella. *Gonyostomum semen* is a smaller plant measuring 40-65 µm long with a prominent, triangular apical depression. There are also numerous small refractive rods (trichocysts) at the periphery of the cells.

PHAEOPHYCEAE OR BROWN ALGAE

This important class of seaweeds is hardly represented in freshwaters. The Brown Algae are characterized by the presence of a brown pigment, fucoxanthin which masks the chlorophyll. The marine forms show great diversity in both their anatomy and reproductive processes.

In Britain, only *Heribaudiella fluviatilis* is at all frequent (Plate 8 Fig. 2). This plant forms thin, dark brown crusts on hard stones, in shaded running water. The cells, which are 4-8 μm wide, develop into close packed tiers up to 80 μm high, sometimes with enlarged, rounded sporangia at their apices. Each cell contains several brown discoid chloroplasts.

THE BACILLARIOPHYCEAE OR DIATOMS

The diatoms are a well defined class of algae characterized by the structure of their silicified cell walls. The wall usually consists of two halves which are virtually identical in shape but sometimes differ in their ornamentation. The two halves fit neatly together like the two halves of a pill- or date box (Plate 16). They are held in this position by intercalary bands (Plate 16m). The silicified wall or *frustule* can usually be observed from two points of view. Looking down on top of the box gives the *valve view* (Plate 16 p) and looking along the edge gives the girdle view (Plate 16 o). Both views are commonly seen under the microscope but the valve view predominates because it is usually the broadest. In a few species, the girdle view predominates and the diatom may require manipulation to make the valve view visible. Some diatoms such as *Amphora* and *Nitzschia* have a peculiar shape and this terminology then no longer applies.

Diatoms may be circular with radial symmetry in valve view (Plate 16 l, p) or boat shaped, with bilateral symmetry (Plate 16 a-i) or they may even be completely asymmetrical. The valve shape and symmetry is an important character in classification. When the two *sides* of the valve are mirror images of each other, they are said to possess *apical symmetry* and when the two ends show mirror image symmetry they possess *transapical symmetry* (Plate 16). Most boat-shaped, or *pennate* diatoms have both apical and transapical symmetry.

Many pennate diatoms are motile and appear to achieve this by secreting mucilage out of a slit-like structure called the *raphe* (Plate 16 i). This may be present on one or both valves and it often divides the valve into two halves. The raphe normally begins close to the valve apex and stops just before reaching the centre of the valve. The regions where it begins and ends are often slightly thickened and called the *polar* and *central nodules* respectively. The raphe is an important character in identification and although narrow it is usually easy to see under high magnification.

The valves are nearly always ornamented with rows of pores, ridges or internal plates which are well seen under the electron microscope. Recent studies have revealed details which the early light microscopists could not interpret accurately and this will probably lead to taxonomic changes in the future. With the light microscope it is usually possible to distinguish four main kinds of ornamentation. These are a) *septa*, b) *ribs* or *costae*, c) *alveoli* and d) *punctate* (Plate 16 c-f). *Septa* consist of long plate-like extensions of the frustule which may almost completely divide the cell into a number of sections like the bulkheads of a ship. *Costae* are prominent wall thickenings which appear as a series of undulations when viewed from the side. *Punctae* are composed of rows of minute perforations and there are many varieties. Sometimes they are covered by a sieve-like structure or they may be isolated slits or pores. *Alveoli* consist of a series of hollow tubes whose upper surface is frequently perforated and sieve-like. In a few genera, the alveoli have regular constrictions and can then be difficult to distinguish from punctae (e.g. *Didymosphenia* Plate 27 Fig. 1). In the pennate diatoms, the median region which often contains the raphe is clear of ornamentation. This region is called the *axial area* (Plate 16 b).

Diatoms are either solitary or colonial plants and the latter may occur as long filaments (*Melosira*) stellate groups (*Synedra*) or zig-zag chains (*Diatoma*). A few forms grow out on long mucilage stalks forming dendroid colonies (*Gomphonema*), or develop in mucilage tubes (*Cymbella*).

The diatoms are a successful class inhabiting all kinds of sites. Some are cold water forms whilst others are terrestrial. There are a number of characteristic planktonic types and others occur on the surface of loose sediments. A huge number of species have

PLATE 16

Structure of the diatom frustule, a-i Cut away view of a pennate diatom, omitting girdle bands and cytoplasm, illustrating features present in different genera. a terminal nodule, b axial area, c various lines of punctae, d alveola, e costa, f septum, g epivalve, h hypovalve, i raphe, j-k Highly magnified sections through frustule showing ornamentation structure in detail, j alveola, k puncta, l-n Exploded view of a centric diatom, l epivalve, m girdle band, n hypovalve, o-r General features: o girdle view of a centric diatom, p valve view, q dorsal margin of a dorsiventral diatom (*Cymbella*), r ventral margin.

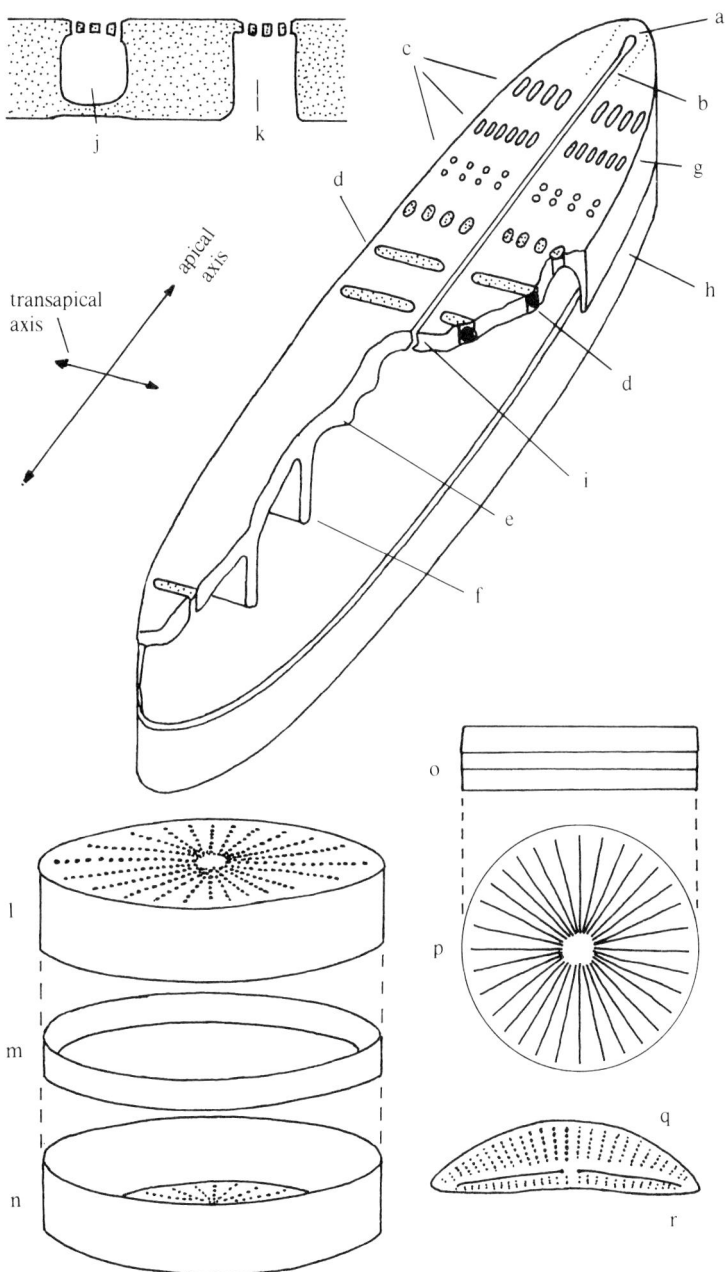

apical axis

transapical axis

a
b
c
d
g
h
d
i
e
f

j
k

l
m
n

o
p
q
r

85

been described on the basis of frustule ornamentation and shape but some distinctive species have been found to show considerable variation within a single stretch of river.

For detailed study, diatoms must be cleaned and mounted so that their fine structure can be observed as clearly as possible. The simplest method is to boil the sample in 10-20 vol hydrogen peroxide for 5-10 minutes and then transfer the suspension to a clean slide where it can either be mounted in water or dried down and mounted in a special resin which increases resolution of details. The number of striations measured in ten micrometres (ITM) is a useful guide in identification. The measurement should be made near the valve centre unless otherwise stated. The striations in some species are so fine that not even the best oil immersion lenses will resolve them.

The diatom cell contains one or more yellow to brownish chloroplasts. Centric species possess a large number of small discoid chloroplasts (e.g. *Melosira*) but pennate diatoms contain either one or two which extend the length of the cell and these may have pyrenoids. The chloroplasts of *Surirella* and similar forms possess two deeply divided and lobed chloroplasts. *Rhopalodia gibba* is remarkable in possessing a cyanobacterium-like inclusion capable of fixing nitrogen.

Diatoms divide in a plane parallel to the valve surface by forming two new valves within the old ones. After separation, the younger valves enlarge slightly to maintain the dimensions of the cell. Sexual reproduction is either oogamous or a form of conjugation. This often results in the production of *auxospores* . These are rounded, masses of protoplasm formed when two diatoms come into close contact and held by a mucilaginous secretion.

Key to Genera

1a Valves disc- or pill-box shaped, usually with radially symmetrical ornamentation. Raphe absent or apparently so, (Centrales).   2

1b Valves fusiform, oval, semicircular, sigmoid or lunate without radially symmetrical ornamentation. Raphe present or absent.   5

2a Cells shortly cylindrical, (0.8)1-2.5(5) times longer than their diameter. Most often seen in girdle view, valves with radiating punctae or smooth. Short marginal spines present in filamentous forms. Cell d. (4)6-40(80) µm. Common, planktonic and benthic.   *Melosira* 1

2b Cells disc-shaped, shorter than their diameter, usually seen only in valve view.   3

3a Central area of the valve smooth, peripheral area with radiating striae or punctae. Valve d. (4)6-13(80) µm, 0.2-0.4(0.6) times as wide as deep. Cells solitary, in groups or in mucilaginous colonies. Common, planktonic and benthic.   *Cyclotella* 2

3b Central area of the valve not differentiated from the peripheral area.   4

4a Valves with 30-50 rows of radiating punctae separated by the same number of smooth rows and with short (1-2 µm) marginal spines. Valves (6)8-40(70) µm in d., ltb 0.3-0.5. Common in lowland lakes and rivers.   *Stephanodiscus* 3

4b Valves evenly ornamented with radiating punctae, 20-70 µm in d., slightly undulate in girdle view. Uncommon.   *Coscinodiscus* 4

5a No raphe present on either valve. (Diatoms belonging to this section include many producing star-shaped, zig-zag, ribbon-like colonies)   6

5b A raphe present on at least one valve. Forms with the raphe obscure at the valve margin usually with the margins appearing to differ in structure from the rest of the valve.   15

6a Frustules acicular or cylindrical, with one or two prominent spines at each apex. Ornamentation consisting of about 10-30 transverse lines interrupted by a lateral zig-zag line. Frustules not separable into two well defined valves, planktonic.   7

6b   Frustules variously shaped but not as above and usually with distinct valve- and girdle views.   8

7a   Frustules acicular with a single spine at each pole, cells (40)60-180(200) µm long, ltb (5)8-20. Scarce.

Rhizosolenia 5

7b   Frustules shortly cylindrical with two parallel spines at each pole, cells (12)20-150 µm long, ltb 4-6. Occasional.

Attheya 6

8a   Frustules with internal septa up to 1 µm or more in thickness, best seen in girdle view. Septa travelling from about half way to right across the cell, normally seen in girdle view. Often centrally inflated in valve view.   9

8b   Septa absent in girdle view.   10

9a   Frustules in valve view with 4-10(20) transverse costae interposed with fine parallel striae. Septa (2)7-8 in girdle view, normally crossing the cell. Solitary or in chains, usually montane. Cells 10-80 µm long, ltb 2-7.

Tetracyclus 7

9b   Frustules with fine parallel striae in valve view, costae absent. Septa 2-7 in girdle view, often only extending half way across the cell. Cells (10)20-70(130) µm long, ltb (valve view) (7)10-25. Common, often forming zig-zag chains.   *Tabellaria* 8

10a   Transverse costae present in valve view.   11

10b   Transverse costae absent.   12

11a   Cells in circular fan-shaped colonies showing the wedge-shaped girdle view. Frustules clavate in valve view with 10-20 often incomplete, transverse costae and fine interposing striae. Cells 15-60(80) µm long, ltb (valve view) 3-10.   *Meridion* 9

11b Cells not in fan-shaped colonies, frustules in valve view fusiform and isopolar with (2)6-25(40) transverse costae with fine interposed striae. Cells (20)30-80(100) μm long, ltb (1.5)3-10(25). Often in zig-zag or ribbon shaped colonies, common, benthic. *Diatoma* 10

12a Valves arcuate and slightly inflated centrally on the ventral side, cells (15)30-150 μm long, ltb 7-10. Occasional in nutrient-poor waters. *Hannaea* 11

12b Valves isopolar. 13

13a Cells forming stellate colonies in girdle view, with all cells in one plane and (4)5-8(20) cells per colony. Frustules elongate with inflated ends in valve view, (30)40-130(150) μm long, (5)8-20(30). Planktonic in lowland lakes. *Asterionella* 12

13b Cells in ribbon-shaped or spangled colonies, common. 14

14a Colonies ribbon-shaped, cells (6)15-40(100) μm long, ltb (1.5)5-20(30). Planktonic or benthic. *Fragilaria* 13

14b Cells in spangled clusters, usually acicular and frequently slightly curved. Cells (5)30-200(500) μm long, ltb (6)10-25(40). Epiphytes, especially in lowland waters. *Synedra* 14

15a Raphe very short and apparent only near the valve ends. Valves approximately bow-shaped, often with inflated ends and without a median axial area. Cells (8)10-100(200) μm long, ltb (3)7-15(25). Common in nutrient poor sites, benthic. *Eunotia* 15

15b Raphe approximately median or lateral, extending to the mid-region of the cell on at least one of the valves. 16

16a Raphe approximately median and present on one valve only. (In *Rhoicosphenia* a very short raphe is present on the other valve). Valves often chevron-shaped in girdle view. 17

16b   Raphe present on both valves, median or lateral, valves not chevron-shaped in girdle view.   19

17a   Valves oval with a small central area and a median raphe. Punctae radiate near the centre, cells (5)10-30(70) µm long, ltb (1)1.2-2.2(2.5). Common epiphytes.   *Cocconeis* 16

17b   Valves elongate-fusiform, common benthic spp.   18

18a   Valves fusiform with a much reduced raphe on the 'pseudoraphe' valve. Girdle view both chevron-shaped and cuneate. Cells 12-75 µm long, ltb 4-10.   *Rhoicosphenia* 17

18b   Valves fusiform, often with swollen apices and a raphe on one valve only and with usually fine striae or punctae. Girdle view not cuneate. Cells (5)8-30(200) µm long, ltb 3-10.   *Achnanthes* 18

19a   Valves with apical and transapical symmetry, raphe median.   20

19b   Valves and raphe without this combination of characters.   28

20a   Raphe bounded on either side by a clear border, 0.15-0.3 times the valve width.   21

20b   Raphe not bounded by such a border.   23

21a   Raphe short, extending inwards from the valve apices 0.2-0.1 times the total valve length. Valves fusiform, (15)20-200(300) µm long, ltb 6-15. Striae very fine, transverse.   *Amphipleura* 19

21b   Raphe extending into the central area of the valve.   22

22a   Valve ornamentation changing when traced from the edge to the axial area. Valves undulate or broadly fusiform, (10)20-70(150) µm long, ltb (2)3-7(9). Benthic.

Diploneis 20

22b   Valve ornamentation the same throughout, consisting of fine tranverse striae. Valves fusiform or rhomboidal, (30)40-90(160) µm long, ltb 6-10. Common, oligotrophic.

Frustulia 21

23a   Valves lined with 15-30 peripheral, cubical chambers, about 0.25 times the valve width. Valves fusiform with fine transverse striae, (20)25-70(100) µm long, ltb 3.5-6. Scarce, base-rich sites. *Mastogloia* 22

23b   Valves without marginal chambers.   24

24a   Valves with one or more lateral interruptions of the fine transverse striae.   25

24b   Valves without lateral interruptions of fine transverse striae *or* valves with coarse punctae or alveoli, the latter sometimes with lateral lineations   *Pinnularia* 27

25a   Central terminations of the raphe hooked or bent in opposite directions. Striae with one to several marginal interruptions and often a differing peripheral ornamentation. Frustules fusiform, sometimes capitate, (20)30-90(200) µm long, ltb (4)6-8. Widespread, benthic.
Neidium 23

25b   Central terminations of the raphe bent to the same side or straight. Ornamentation with 1-2 linear interruptions. Valves fusiform, sometimes gibbous or with capitate apices, (20)30-100(200) µm long, ltb (3)4-7(9). Widespread, benthic.   *Caloneis* 24

26a   Striae alveolar and rib-like. Raphe often twisted and appearing complex. Frustules solitary or in short ribbons, elongate-fusiform (15)30-120(250) µm long, ltb (4)5-8(10). Common, base-poor areas.   *Pinnularia* 25

26b   Striae not alveolate, raphe usually simple.   27

27a   Valves with a thickened central area without ornamentation and often with apical thickenings. Cells fusiform, sometimes gibbous or capitate, (15)20-100(200) µm long, ltb (5)6-8(10). Benthic.   *Stauroneis* 26

27b   Valves without a thickened central area. Frustules elongate-ellipsoid or fusiform, sometimes gibbous or capitate, (8)15-60(120) µm long, ltb (2)4-8(10). Common, benthic.   *Navicula* 27

28a   Frustules ovate, composed of two lunate valves joined along their straight ventral edges. Raphe appearing as a double set of lines parallel to the median partition. Frustules (6)10-40(80) µm long, ltb 1.8-2.5. Frequent, base-rich sites. *Amphora* 30

28b   Frustules otherwise.   29

29a   Raphe situated upon or close to a valve ridge. Raphe often with a beaded appearance, and difficult to see.   33

29b   Raphe median or lateral, without a beaded appearance and not upon a ridge, generally easy to see.   30

30a   Valves sigmoid, with neither apical nor transapical symmetry. Striae fine, transverse, raphe median and sigmoid. Valves (40)80-180(220) µm long, ltb 8-12. Benthic. *Gyrosigma* 28

30b   Valves with either apical *or* transapical symmetry.   31

31a   Valves with transapical symmetry, lunate or crescent-shaped with the raphe often slightly off-centre and parallel to the ventral margin. Valves (12)20-70(200) µm long, ltb (4)5-8(10). Striae usually radiating. Common, benthic. *Cymbella* 29

31b   Valves with apical symmetry, cuneate in girdle view.   32

32a   Valves usually with an isolated puncta in the central area, with fine striae. Valves often capitate, with the two apices differing in size or shape, (15)20-60(80) µm long, ltb (4)5-8(12). Epiphytic or benthic. *Gomphonema* 31

32b   Valves large with 2-5 isolated pores in the central area and ornamented with radiating coarsely punctate alveoli. Valves capitate, with one apex less swollen than the other. 50-135 µm long, ltb 8-10. Neutral or basic streams. *Didymosphenia* 32

33a   Frustules in valve view arcuate or crescent-shaped, with transverse costae, epiphytic.   34

33b Without this combination of characters. 35

34a Raphe in valve view, close to the dorsal margin throughout and with 15-60 transverse costae and intervening striae. Valves (20)35-200(300) µm long, ltb (3)5-8. Usually seen in the gibbous girdle view. *Rhopalodia* 33

34b Raphe in valve view bow-shaped, meeting the valve edge at the apices. Valves semicircular or arcuate, sometimes capitate, (15)25-100(200) µm long, ltb (4)6-11, transverse costae 10-50. *Epithemia* 34

35a Costae absent, benthic. 36

35b Costae present 37

36a Frustules with a strong sigmoid keel, lanceolate in valve view, broadly winged in girdle view, (35)50-200(350) µm long, 1.8-3.2 ltb (girdle view), uncommon. *Entomoneis* 40

36b Valves lanceolate, fusiform or sigmoid with a linear keel and a series of dots extending along the raphe margin, valves (10)20-150(600) µm long, (4)6-25(30) ltb, common. *Nitzschia* 36

37a Valves with broad (1-5 µm) costae near the margins containing the raphe, costae disappearing towards the axial area. 38

37b Valves with 6-25 entire transverse costae and intervening striae, fusiform or slightly asymmetric, (6)10-40(60) µm long, ltb (2)4-10. Raphe generally marginal but variable in position. Benthic. *Denticula* 35

38a Valves strongly flattened in girdle view. 39

38b Valves not flattened but fusiform or clavate, often large. Ribs thick with fine intervening striae, valves (20)35-150(400) µm long, ltb (1.8)4-6(10). Benthic. *Surirella* 37

39a Valves almost circular, saddle-shaped in girdle view, costae radiating, 60-200 µm d., uncommon. *Campylodiscus* 39

39b Valves elongate-ellipsoid, often capitate, with regular
     transverse undulations and fine striae, ribs confined to the
     valve margin. Valves (30)50-150(230) µm long, ltb 1.2-5.
     Benthic.   *Cymatopleura* 38

The Centric Diatoms
1. Melosira

Frustules cylindrical and often in long chains where the girdle view is exposed. The valve view is rarely seen since the cells tend to topple over on their sides. The species are often extremely abundant especially among aquatic mosses in lowland lakes and rivers and in the plankton of eutrophic lakes. (Id: ornamentation, presence or absence of short spines connecting the cells together; 15).

M. varians (Plate 17 Fig. 1) has almost plain unornamented frustules which measure 8-35 µm wide, ltb 1.2-1.6 in girdle view. It is often common in still waters.

M. granulata (Plate 17 Fig. 2) has coarsely punctate frustules 5-20 µm wide, ltb 1.6-2 in girdle view. There are about 10 irregular rows of punctae visible. *M. italica* is a frequent planktonic species with about 10 short connecting spines visible in girdle view, and the wall is finely punctate. The cells measure (4)6-25 µm wide, ltb 2.2-2.7. (Plate 17 Fig. 3). Several species occur as narrow varieties with helically or spirally twisted filaments.

2. Cyclotella

Cells solitary or in short chains, pill-box shaped, often seen in girdle view. These small diatoms are widely distributed, particularly in the plankton. (Id: ornamentation in valve view; 30).

C. meneghениana is 10-30 µm wide with 40-50 rows of radiating striae (Plate 17 Fig. 4) and *C. kuetzingiana* is 10-40 µm wide with about 90 radiating striae (Plate 17 Fig. 5). Both species are common and have a punctate central area.

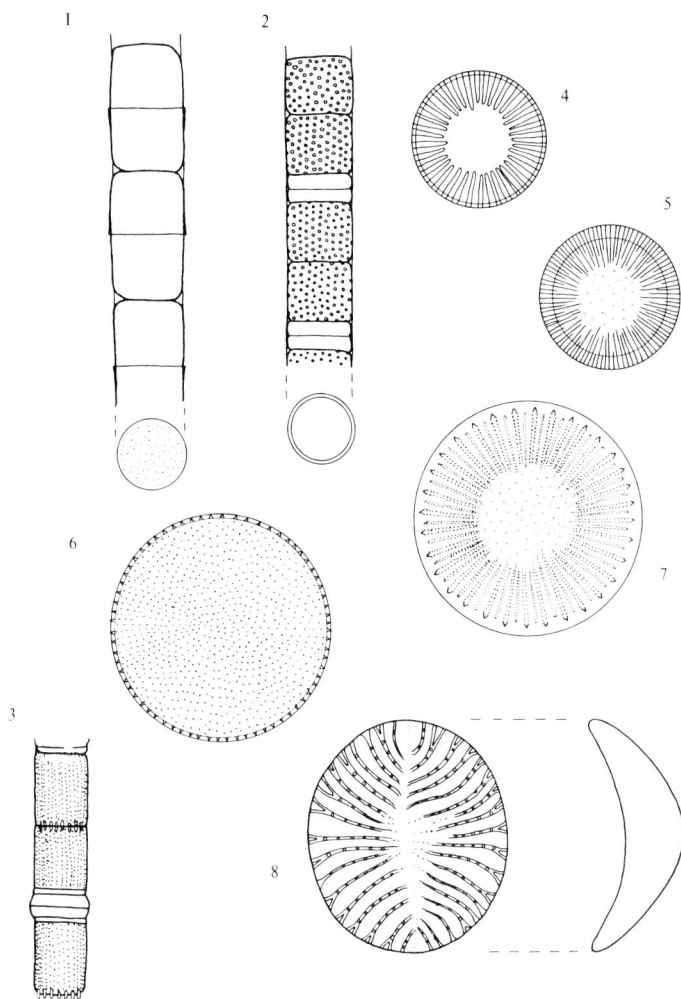

PLATE 17

1 *Melosira varians*, girdle and valve views, x 750, 2 *M. granulata*, girdle and valve
views, x 750, 3 *M. italica*, girdle view x 750, 4 *Cyclotella menegheniana* valve view,
x 1350, 5 *C. kuetzingiana* x 950, 6 *Coscinodiscus lacustris*, x 900, 7 *Stephanodiscus
hantzschii*, x 3000, 8 *Campylodiscus noricus*, x 450, British Museum Collection.

3. Stephanodiscus

Valves with alternating rows of punctae and clear areas of about equal size, together with a series of short marginal spines. (Id: shape and size of marginal spines, ornamentation; 10).

S. hantzschii (Plate 17 Fig. 7) is sometimes abundant in lowland lakes and rivers. The valves are (8)10-20 µm wide with fine dispersed punctae.

4. Coscinodiscus

Valves with equally spaced, radiating punctae and a narrow, differentiated margin. (Id: presence or absence of spines, ornamentation; 4).

C. lacustris (Plate 17 Fig. 6) is an uncommon diatom with minute marginal spines and fine radiating punctae, with valves 25-75 µm wide.

5. Rhizosolenia

Frustules cylindrical with a long or short, asymmetric spine at each pole and a series of transverse bands crossed by a zig-zag girdle. (Id: shape, number of bands; 3).

R. eriensis is the best known species 40-150 µm in length, ltb (including spines) (4)10-20. With the spines accounting for one half or less of the total cell length (Plate 18 Fig. 7). The species are planktonic, but uncommon.

6. Attheya

Frustule cylindrical, with two parallel spines at each apex, giving an H-shaped cell. The degree of mineralization is often low so the two species are easily overlooked in the plankton.

A. lata is almost square in shape and 10-20 µm in length excluding the spines. *A. zachariasii* is more elongate, ltb 1.6-4 and 20-40 µm long excluding spines. Rare in lowland lakes. (Plate 18 Fig. 6).

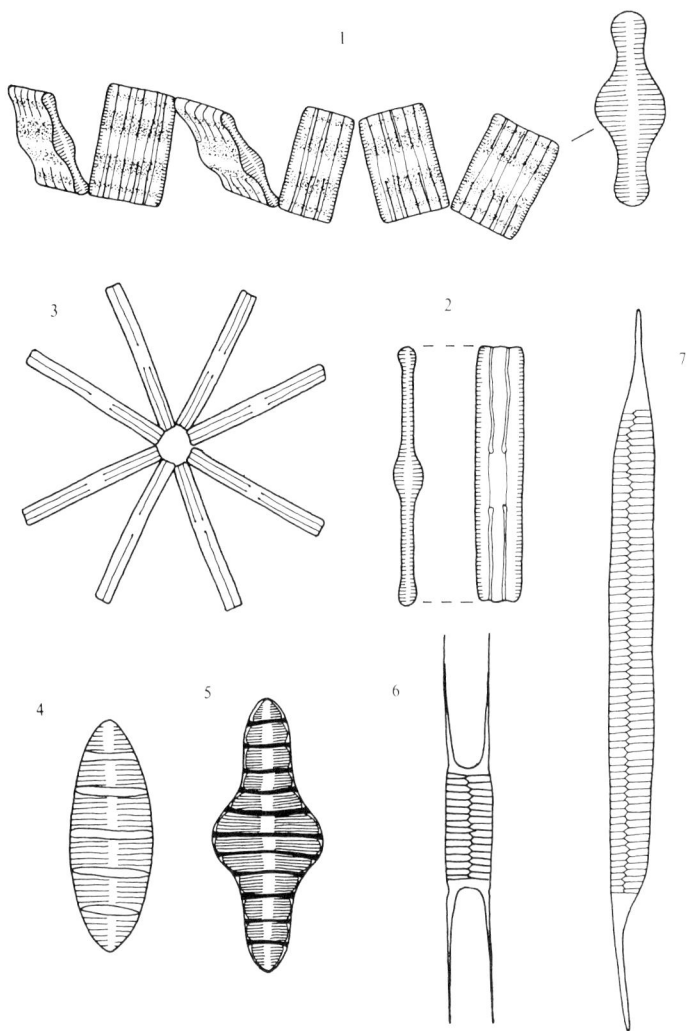

PLATE 18

1 *Tabellaria flocculosa*. Chain of cells showing mainly girdle view x 500, valve view x 950, 2 *T. fenestrata* Valve and girdle view x 800, 3 *T. fenestrata* var. *asterionelloides* x 600, 4 *Tetracyclus rupestris* x 2250, 5 *T. lacustris* x 1100, 6 *Attheya zachariasii*, Tai Hu, China, x 600, 7 *Rhizosolenia eriensis* x 1500, British Museum Collection.

The Pennate Diatoms

7. Tetracyclus

Frustules with septa showing in girdle view and transverse costae in valve view. Two species are widely distributed but uncommon in Britain, and normally seen only in girdle view.

T. lacustris (Plate 18 Fig. 5) has 10-20 costae with valves 30-80 µm long, ltb 2-7 and is centrally gibbous. *T. rupestris* has lanceolate valves (Plate 18 Fig. 4) 10-30 µm long, ltb 4 with just 4-5 costae.

8. Tabellaria

Frustules forming long zig-zag or stellate colonies, usually seen in girdle view. The large septa are easily seen.

There are two species, *T. flocculosa* with valves 10-100 µm long, ltb 2-6 and 8-20 septa is almost ubiquitous in Britain among aquatic plants (Plate 18 Fig. 1). *T. fenestrata* is a more slender species with valves 30-120 µm long, ltb 6-10 and 4-8 septa visible in girdle view (Plate 18 Fig. 2) and is less common. The var *asterionelloides* (Plate 18 Fig. 3) forms star-shaped colonies and is common in the larger oligotrophic lakes of Britain. Some forms can be confused with *Asterionella*.

9. Meridion

The fan-shaped colonies of *M. circulare* are unlikely to be mistaken with any other species. The wedge-shaped frustules are seen mainly in girdle view and the costae appear as a series of small checks along the walls (Plate 19 Fig. 1). The cells measure 12-80 µm long, ltb 2.2-7 in valve view. This species is often abundant in slow flowing streams and ditches.

PLATE 19

1 *Meridion circulare* Colony in girdle view x 1150, valve view x 1400, 2 *Diatoma vulgare*, colony in girdle view and valve view x 900, 3 *D. hiemale* girdle and valve view x 1000, 4 *D. hiemale* var. *mesodon* girdle and valve view x 750, 5 *Hannaea arcus* Ben Lui, Argyll, x 1400, 6 *Fragilaria capucina*, valve and girdle view x 1100, 7 *F. crotonensis* valve and girdle view, Windermere, Cumbria, x 700

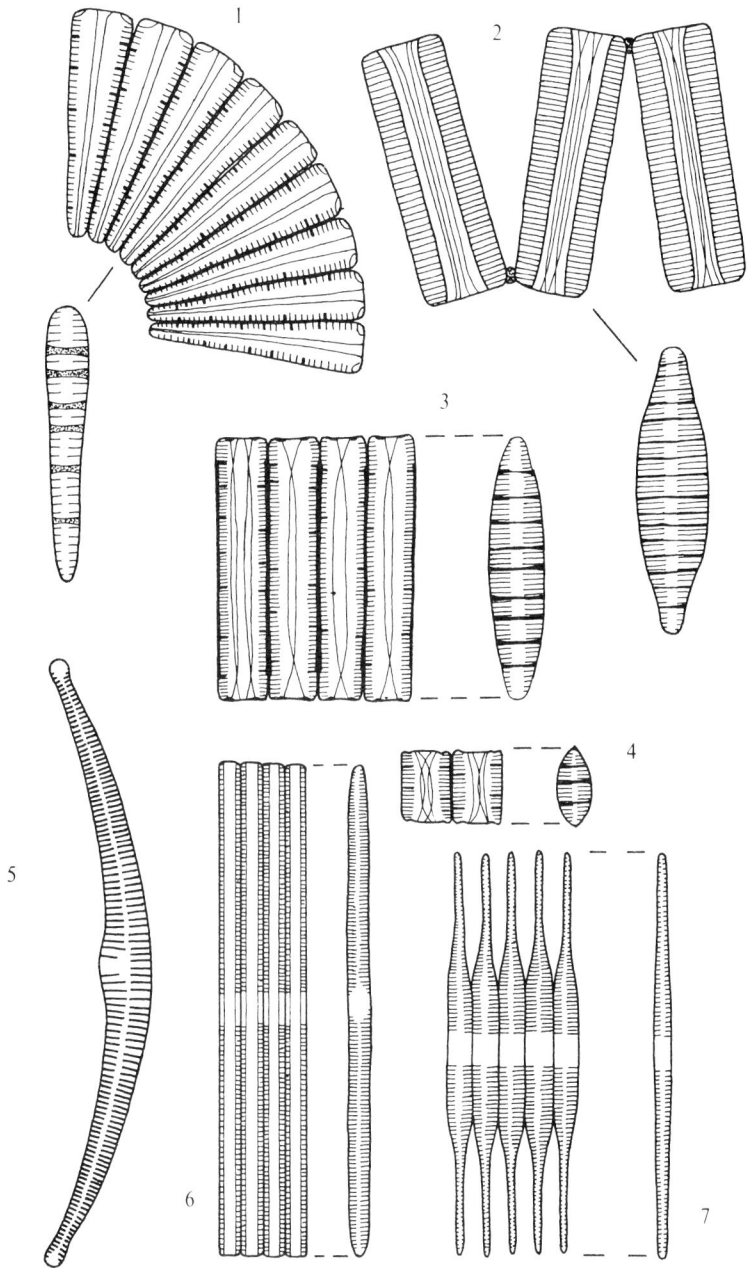

10. Diatoma

Frustules often in ribbon-shaped or zig-zag colonies, rectangular in girdle view and with several transverse costae which can be seen as a series of small points at the cell margin. (Id: shape and valve ltb; 7).

There are two common forms in Britain which are often abundant among aquatic plants in hard water. *D. vulgare* forms zig-zag colonies with cells 30-60 µm long, ltb 3-5 in valve view and 10-20 costae (Plate 19 Fig. 2). *D. hiemale* forms short ribbons of cells which are 30-100 µm long, valve ltb 4-8 with 10-20 costae (Plate 19 Fig. 3). The small variety *mesodon* (Plate 19 Fig. 4) has smaller frustules, 10-30 µm long and 1-5 costae. It is almost square in girdle view and locally common.

11. Hannaea

Valves bow-shaped with a central inflation on the dorsal margin and occasionally with slightly capitate apices (Syn. *Ceratoneis*).

H. arcus (Plate 19 Fig. 5) is the only species with cells 15-150 µm long, ltb 3-9. The fine striae are almost parallel and 14-14 ITM. This diatom is sometimes frequent in cool, base-poor waters.

12. Asterionella

Frustules in girdle view, forming cartwheel-shaped colonies containing (4)6-8(15) cells. In valve view the frustules have slightly capitate apices and very fine parallel striae 20-30 ITM.

A. formosa is a common and well studied planktonic diatom which may be found in all kinds of water but is most abundant in mesotrophic and eutrophic lakes. In common with other bloom-forming diatoms, *Asterionella* gradually removes silica from the water in which it grows until no more is available for growth. Once this occurs, the crop declines rapidly, in late spring or early summer. The cells measure 40-130 µm long, valve ltb 8-20 (Plate 28 Fig. 1).

13. Fragilaria

Frustules joined into ribbons by the valve faces. Valves fusiform, sometimes centrally gibbous or with capitate ends but usually seen only in girdle view in uncleaned material. The striae are fine and usually parallel. The species are abundant, both in the plankton and among submerged plants and stones. (Id: valve shape, striae position and number; 30).

Key to the commoner British species

1a   Cells attached by the valve centres only, appearing comb-like. Valves linear or slightly capitate, 40-170 μm long, ltb 15-25. Striae 15-18 ITM, central area rectangular. Common phytoplankter in mesotrophic lakes.

F. crotonensis (Plate 19 Fig. 7).

1b   Cells joined by their entire faces.   2

2a   Valves broadly gibbous, 7-25 μm long, ltb 2-4. Striae 13-18 ITM, acute near the central area. A variable planktonic and benthic species.   *F. construens*
(Plate 20 Fig. 2)

2b   Valves linear, if gibbous then swollen less than twice the valve width near the apices.   3

3a   Valves slightly inflated on one side, 10-40 μm long, ltb 10-15. Striae parallel, 12-16 ITM. Widespread, benthic.

F. vaucheriae (Plate 20 Fig. 1)

3b   Valves symmetrical, benthic.   4

4a   Striae 7-12 ITM, obtuse at apices and crossed by two fine lines. Valves 3-25 μm long, ltb 2-4, rectangular or square in girdle view.   *F. pinnata* (Plate 20 Fig. 3)

4b   Striae 12-19 ITM   5

5a   Central area rectangular, almost reaching valve margin. Valves (25)40-100(150) μm long, ltb 20-30. Striae parallel.

F. capucina (Plate 19 Fig. 6).

5b   Central area absent, valves (10)12-120 μm long, ltb 10-15.

F. virescens (Plate 20 Fig. 4)

14. Synedra

Cells in irregular bundles or in twos joined side by side but not forming ribbons. Valves needle-shaped or fusiform, often capitate. In cleaned material, the frustules are difficult to separate from *Fragilaria*. (Id: valve shape, axial area, striae number; 30).

Only two of the several common forms are illustrated. *S. ulna* may be distinguished by its rostrate or wedge-shaped apices and the quadrate, unswollen central area. The valves measure (50)75-150(350) µm long, ltb 15-45 and are gradually attenuated towards the apices. The striae are parallel and 8-12 ITM (Plate 20 Fig. 5). *S. capitata* (Plate 20 Fig. 6) has capitate and wedge shaped apices without a central area. The valves are 125-500 µm long ltb 15-25 and striae 8-11 ITM. The species are common epiphytes in lowland waters.

15. Eunotia

Cells isolated or in filaments, valves arcuate with a short raphe which is easily overlooked, situated on the ventral edge near the apices. The striae are generally parallel and traverse the entire valve. (Id: valve shape, striae numbers; 40).

The three illustrated species are *E. pectinalis* (Plate 20 Fig. 11) with weakly arcuate valves narrowed on either side of the central area, measuring 17-140 µm long, ltb 7-12, striae 7-12 ITM. In some varieties the dorsal margin is slightly undulate. *E. curvata* is a smoothly bow-shaped form with the apices gradually attenuated, 20-150 µm long, ltb 6-25 and striae 13-18 ITM (Plate 20 Fig. 12). *E. arcus* which has strongly capitate apices and broadly rounded ends, is 17-90 µm long, ltb 5-11 with striae 8-14 ITM (Plate 20 Fig. 10). Species of *Eunotia* are most often seen in nutrient-poor waters, often among *Sphagnum*.

PLATE 20

1 *Fragilaria vaucheriae* Glen Coe, x 1500, 2 *F. construens* x 2000, 3 *F. pinnata* Malham Tarn, Yorks. x 2250, 4 *F. virescens* x 1200, 5 *Synedra ulna* colony x 600, 6 *S. capitata* x 350, 7 *Cocconeis placentula* x 750, 8 *C. pediculus* x 1700, 9 *Rhoicosphenia curvata* British Museum Collection, valve and girdle view x 1500, 10 *Eunotia arcus* Lock Seil, Argyll. x 1500, 11 *E. pectinalis* x 900, 12 *E. curvata* x 850

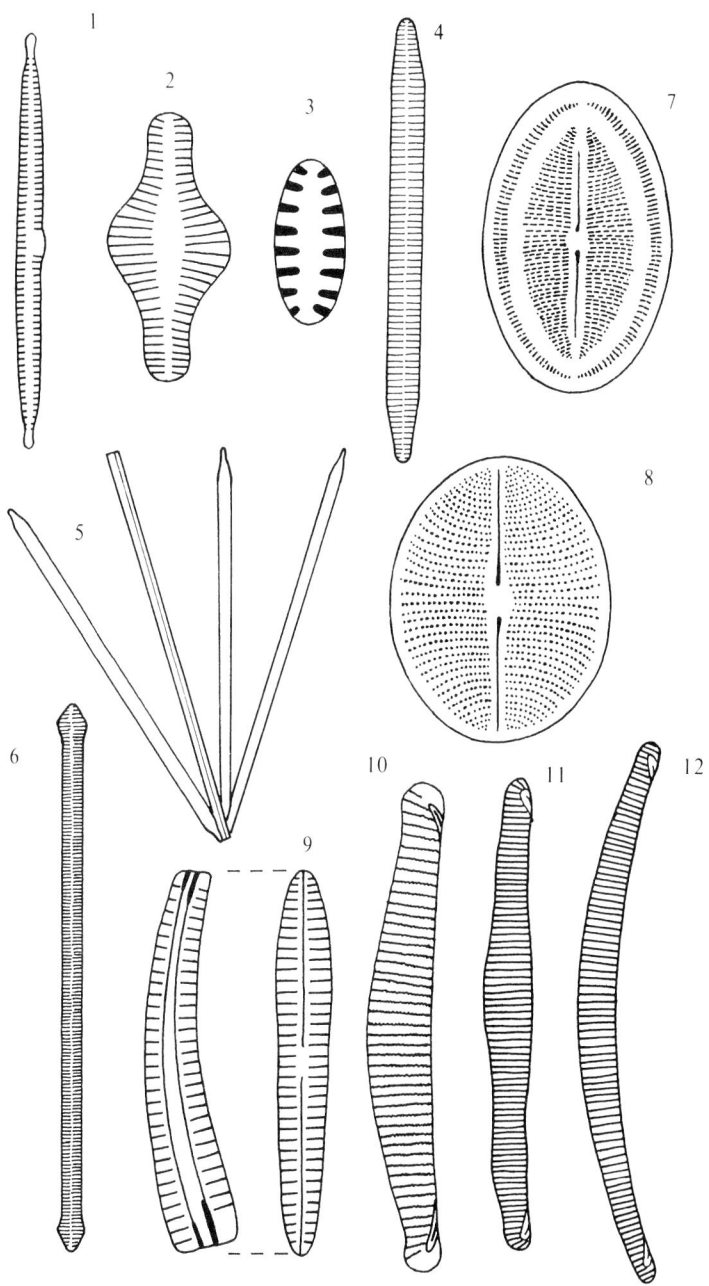

16. Cocconeis

The two common species could easily be mistaken for a centric diatom but careful examination will reveal a linear raphe on one of the valves. The species are often attached closely to the stems and leaves of aquatic plants and they are very common.

C. pediculus (Plate 20 Fig. 8) has a clear marginal area 1-2 μm wide on the raphe valve. The valves measure 11-30(45) μm long, ltb 1.1-1.4 *C. placentula* (Plate 20 Fig. 7) differs by having two clear marginal areas separated by a small number of punctae. The valves are 10-70 μm long, ltb 1.4-1.7(2.1) which are less strongly arched than those of the preceding. Both species have a small central area and curved rows of punctae.

17. Rhoicosphenia

The single species, *R. curvata* (Plate 20 Fig. 9) is well characterized by its bent, wedge-shaped girdle view. The valves are fusiform or club-shaped with a narrow axial area. The valves differ in structure since a normal raphe appears on one and a much reduced raphe on the other, 12-75 μm long, ltb 4-10 with slightly radiate striae 11-13 ITM. The frustules are commonly seen among other algae and mosses in base-rich waters.

18. Achnanthes

Species belonging to this group are common benthic diatoms which are chevron-shaped in girdle view. Careful examination is required because the two halves differ both in ornamentation and the presence or absence of a raphe. The valve without a raphe is called the *pseudoraphe* valve. (Id: valve shape, position, form and number of striae; 40).

PLATE 21

1 *Achnanthes exigua*, two valve views and girdle view x 4000, 2 *A. lanceolata* Malham Tarn, Yorks., pseudoraphe valve x 3000, 3 *A. linearis* raphe view x 3750, 4 *A. inflata* x 1300, 5 *A. microcephala* x 2800, 6 *A. minutissima* two valve views x 4200, 7 *Amphipleura pellucida* valve view x 750 and ornamentation x 4000, 8 *Diploneis elliptica* x 750, 9 *Frustulia rhomboides* x 1000

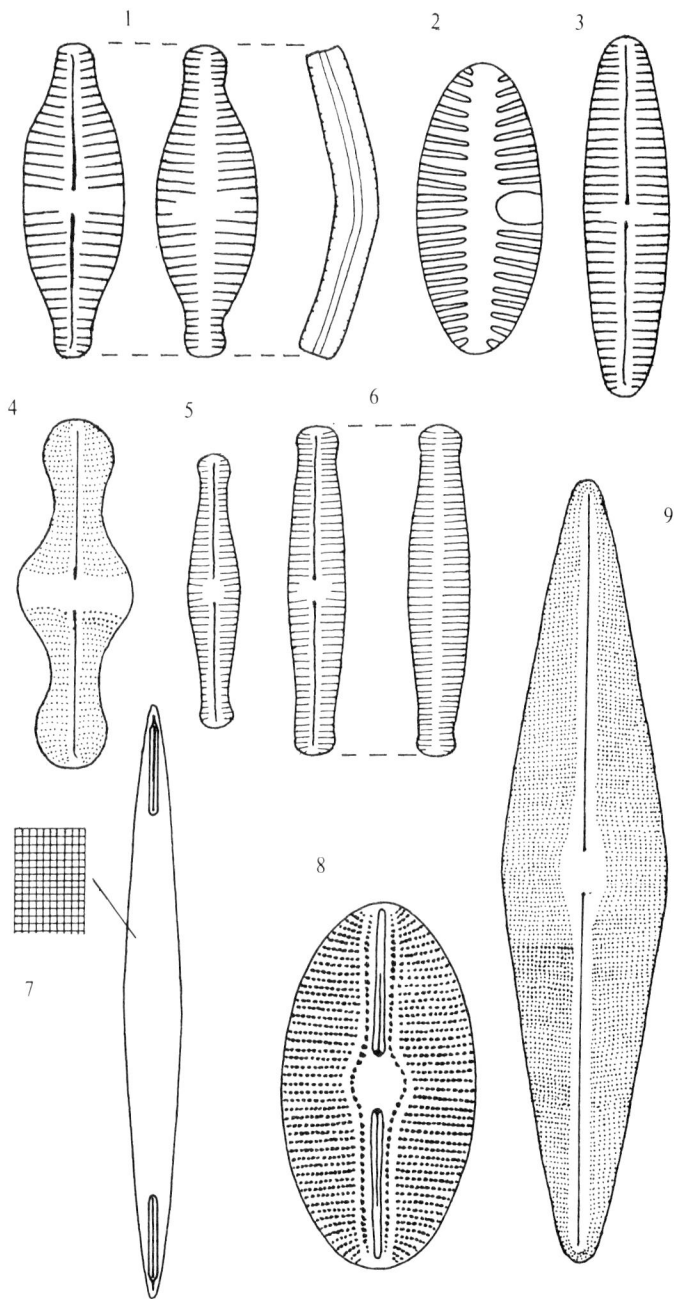

Key to common British species

1a   Pseudoraphe valve with an arched structure centrally placed on one side. Valve shape variable, striae slightly obtuse, 11-14 ITM. Valves 12-30 µm long, ltb (2)3-4(5).
A. lanceolata (Plate 21 Fig. 2)

1b   Arched structure absent.   2

2a   Striae on both valves distinctly punctate, parallel or slightly curved, 9-13 ITM. Valves capitate or gibbous, 30-65 µm long, ltb 2.5-5.   *A. inflata* (Plate 21 Fig. 4)

2b   Striae linear, not distinctly punctate (a group of common forms which are difficult to separate).

3a   Striae near the pseudoraphe valve apices not more than 25 ITM. Valves capitate, 7-17 µm long, ltb 2.5-3.5 with the central area extending to the margin in the raphe valve. *A. exigua* (Plate 21 Fig. 1)

3b   Striae near pseudoraphe valve apices about 30 ITM.   4

4a   Central area of raphe valve extending to the valve edge, striae 24-26 ITM on this valve.   5

4b   Central area of both valves small or absent.   6

5a   Striae slightly acute near the central area. Valves ovate-fusiform 10-20 µm long, ltb 4-6.   *A. affinis*

5b   Striae parallel, valves ovate-fusiform, sometimes capitate, 10-20 µm long, ltb 4-6.   *A. linearis* (Plate 21 Fig. 3)

6a   Striae on pseudoraphe valve 30 ITM, acute near the central area. Valves ovate-fusiform, 5-20(40) µm long, ltb 4-6(8).
A. minutissima (Plate 21 Fig. 6)

6b   Striae on pseudoraphe valve 26 ITM, acute near central area. Valves ovate-fusiform, often capitate, 8-26 µm long, ltb 4-6(8).   *A. microcephala* (Plate 21 Fig. 5)

19. Amphipleura

This is a genus of diatoms with extremely fine rows of punctae (25-40 ITM) and a short raphe enclosed by a silica rib which extends no less then half way towards the central area. (Id: raphe structure, striae number, valve shape; 2).

A. pellucida (Plate 21 Fig. 7) has fusiform valves 80-140 µm long, ltb 11-15 and fine transverse striae 37-40 ITM, only resolved by the best lenses. It is widespread in Britain, with a preference for hard water.

20. Diploneis

The raphe of these species is placed within a thick silica ridge and the ornamentation adjacent to the ridge differs from that on the rest of the valve. (Id: number and structure of costae, valve shape; 20).

D. elliptica (Plate 21 Fig. 8) has broadly elliptic valves and a row of pores surrounding the axial rib. The valve is crossed with costae separated by rows of alveoli although the former are not very clear in this species. Valves 20-130 µm long, ltb 1.7-2.2. Occasional in lakes and bogs.

21. Frustulia

Valves usually rhombic with rounded edges and a prominent raphe within a thickened silica ridge. The striae are fine and parallel. (Id: Raphe structure, ornamentation; 5).

F. rhomboides is a common species of mountain bogs and streams, with valves 70-160 µm long, ltb 4-6 and striae 20-30 ITM (Plate 21 Fig. 9).

22. Mastogloia

Valves elliptical or lanceolate with distinct marginal chambers extending along both sides. Striae parallel or slightly radiate, coarsely or finely punctate with a narrow axial area. (Id: raphe structure, central area, ornamentation; 4).

Most species seem to be confined to brackish water but *M. smithii* appears to be a freshwater species. It has elliptical valves

20-45 µm long, ltb 3.5-4.5 and punctate striae 18-19 ITM and 6-8 chambers ITM. None of the species common. (Plate 22 Fig. 3).

23. Neidium

Valves elliptical or fusiform with rounded or capitate ends. The axial area is narrow and the valves are ornamented with rows of punctate striae which are interrupted by a series of small gaps near the margin giving a regular, bead-like appearance. (Id: raphe structure, valve shape, ornamentation; 20).

N. iridis (Plate 22 Fig. 4) is a handsome species frequently found in lakes and bogs of all kinds. The valves are 50-200 µm long, ltb 4.5-6 with 14-18 striae ITM and 2-3 interruptions.

24. Caloneis

Valves elliptical to fusiform with variable axial and central areas. The striae are fine and alveolate, rather like those found in *Pinnularia* and they have a series of fine longitudinal openings which appear as a series of lines. (Id: shape of axial area, valve shape, ornamentation; 30).

C. bacillum (Plate 22 Fig. 2) is frequently found on rocks in streams. The valves are elliptical, 15-45 µm long, ltb 3.5-4.5 with a broad, lanceolate axial area extending to the valve margin. The striae are 24-30 ITM and there is a single longitudinal line near the margin. *C. ventricosa* (Plate 22 Fig. 1) is another widely distributed species with slightly wedge-shaped and capitate apices and a swollen mid-region. The cells measure 20-85 µm long, 3.5-6 ltb with parallel striae 16-22 ITM. There is a single marginal lineation and a narrow axial area.

25. Pinnularia

Valves elliptical, linear or fusiform, often large and usually with rounded apices. The striae are alveolate and non-punctate

---

PLATE 22

1 *Caloneis ventricosa* x 2300, 2 *C. bacillum* x 1800, 3 *Mastogloia smithii* x 2200, 4 *Neidium iridis* British Museum Collection, x 750, 5 *Surirella biseriata* x 500, 6 *Cymatopleura solea* Lewes, Sussex. x 580, 7 *Denticula tenuis* x 1300

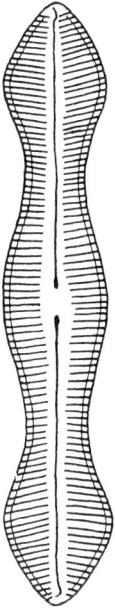

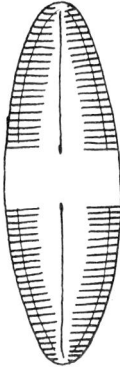

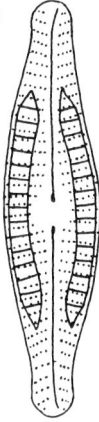

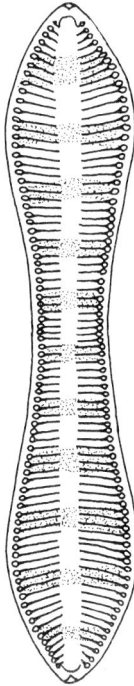

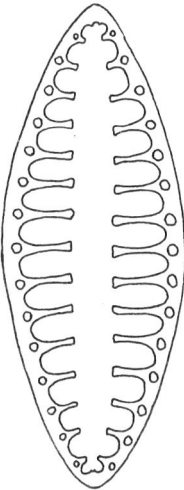

but a series of interior openings are sometimes apparent as 1-several lines following the valve margins. (Id: valve shape, ornamentation, presence of longitudinal lines; 200). The genus is divided into six sections and a species from each section is described below. These algae are often abundant in nutrient-poor waters.

1a   Axial area occupying more than half the cell width in the midvalve region (Brevistriatae). *P. brevistriata* (Plate 23 Fig. 2) with short, parallel striae 7-10 ITM and linear valves 70-135 µm long, ltb 6-8.

1b   Axial area narrow.   2

2a   Striae much less than 6 ITM (Distantes). *P. lata* (Plate 23 Fig. 3) with linear, non-capitate valves 70-160 µm long, ltb 6-7. Striae 2.5-5 ITM, acute near the central area. Axial area 0.25-0.3 times valve width.

2b   Striae § 6 ITM.   3

3a   Striae crossed by fine longitudinal bands   4

3b   Striae without bands.   5

4a   Raphe filamentous (Major). *P. major*, a large diatom with linear valves which are sometimes centrally gibbous, 140-200 µm long, ltb 6-8. Striae 5-7 ITM with 1-2 longitudinal bands, axial area narrow. (Plate 23 Fig. 7)

4b   Raphe with a twisted appearance (Pinnularia). *P. viridis* (Plate 23 Fig. 6). Valves linear, 50-170 µm long, ltb 5-7 with coarse striae 6-9 ITM and 1-2 longitudinal bands. The narrow axial area, 0.2 times the valve width distinguishes this species from most of the others in the section.

5a   Valves capitate, < 75 µm long. (Capitatae). *P. biceps* (Plate 23 Fig. 4). Valves 30-80 µm, ltb 3-5. The central area is rhomboidal and the axial area 0.2-0.3 times the valve width. Striae acute near the central area and obtuse near the apices, 9-14 ITM near the apices.

5b   Valves not capitate, frequently > 75 µm, striae acute near the axial region. (Divergentes). *P. divergens* has a rounded central area with two adjacent rounded thickenings. The axial area is 0.25-0.3 times the valve width, with striae 9-12 ITM. Valves 50-140 µm long, 6-7 ltb (Plate 23 Fig. 5).

26. Stauroneis

Valves elliptical or lanceolate, with rounded or capitate ends. Axial area narrow with the central area broad and thickened, extending to the valve margins. Some species possess apical thickenings which appear as a convex line and give the apex a double-image appearance. The species are common on rocks and aquatic plants, perhaps mainly in oligotrophic waters. (Id: valve shape and structure, ornamentation; 40). Some species are difficult to distinguish from *Navicula*.

S. phoenicicentron (Plate 23 Fig. 1) is widely distributed with lanceolate valves, 70-380 µm long, ltb 5-6 with rounded apices. It lacks apical thickenings and the striae are punctate, 12-17 ITM.

27. Navicula

This large genus includes many common benthic diatoms and keys are provided below to six of the 14 or so sections recognized by diatom experts. Some common or typical species are described under each section. (Id: valve shape, ornamentation and position of striae, valve structure; 300). The species occur in all kinds of aquatic habitats.

1a	Raphe lying within a silica thickening which expands laterally near the valve apices.	Sect. Bacillum (1)
1b	Raphe not within a thickening.	2
2a	Striae clearly resolved as separate punctae (a good x 100 objective lens is required for this).	3
2b	Striae not resolvable into separate punctae	5
3a	Punctae arranged in both longitudinal and transverse lines.	4

3b	Punctae arranged into transverse lines, approximately equidistant and radiate near the central area.
	Sect. Punctulata (4).

4a	The two lines of punctae at right angles to each other.
	Sect. Cuspidata (2)

4b	Longitudinal lines oblique to the transverse lines.
	Sect. Decussata (3)

5a	Striae with a regular, barred structure (Plate 24 Fig. 1).
	Sect. Navicula (6)

5b	Striae otherwise, very fine and radiate near the central area, axial area narrow.
	Sect. Minuscula (5)

Sect. 1 Bacillum

N. pupula (Plate 23 Fig. 8) is a frequent species with the terminal nodules reaching the valve margins. Valves linear-lanceolate, 20-40 µm long, ltb 3.5-4.5 with finely radiate and punctate striae 13-17 ITM. The smaller species require careful examination to detect the thickening around the raphe.

Sect. 2 Cuspidata

A small section with one common species, *N. cuspidata* (Plate 24 Fig. 6). Valves lanceolate, 30-120 µm long, ltb 3.6-4 with small axial and central areas. Striae forming regular transverse and longitudinal lines, 14-24 ITM. Some forms have a faint locular structure reminiscent of *Mastogloia*.

Sect. 3 Decussata

A section to the last but the striae have a pronounced oblique pattern across the valve, but this is not always evident through the entire valve.

PLATE 23

1 *Stauroneis phoenicicentron* x 450, 2 *Pinnularia brevistriata* x 900, 3 *P. lata* x 600, 4 *P. biceps* x 1800, 5 *P. divergens* x 850, 6 *P. viridis* x 1000, 7 *P. major* x 750, 8 *Navicula pupula* x 2000

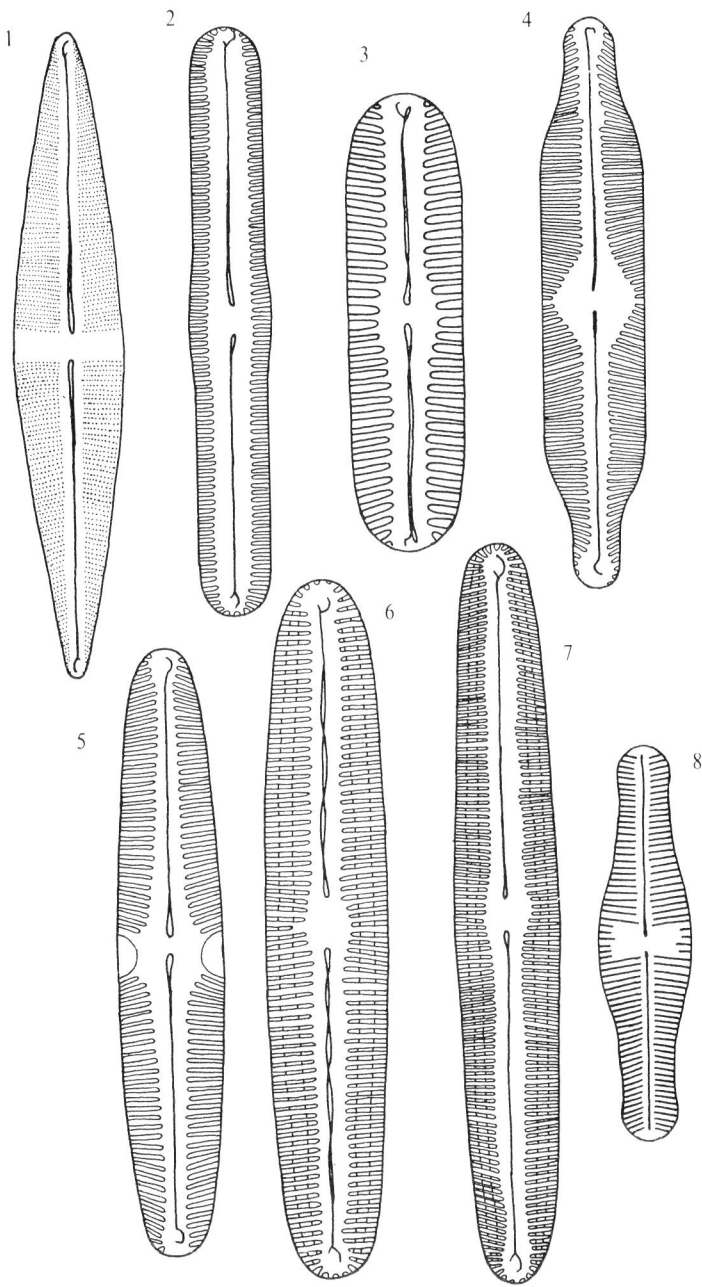

N. placenta (Plate 24 Fig. 7) is a species with lanceolate valves 35-45 μm long, ltb 2.5-3 and striae 22-27 ITM. The median ends of the raphe are set apart from each other and this species is sometimes found among terrestrial mosses.

Sect. 4 Punctulata

The section includes species with clearly punctate and more or less radiate striae which occasionally form irregular longitudinal lines.

N. mutica (Plate 24 Fig. 8) is distinguished by the presence of an isolated punctum present on one side of the central area. The valves are lanceolate with rounded ends, 10-40 μm long, ltb 2.5-3.5 and the radiate striae are 14-20 ITM. Widespread, sometimes terrestrial.

Sect. 5 Minuscula

A fairly large section including mostly small species with a narrow axial area.

A widely distributed species is *N. minuscula* (Plate 24 Fig. 9) with fine, radiate striae 30-34 ITM and elliptical-lanceolate valves 10-15 μm long, ltb 2.5-3. The axial area is narrow and the central area indistinct.

Sect. 6 Navicula

This is the largest section containing a huge number of species. Only a few of the best known types are described here. The species are distinguished mainly by the angle of the apical striae, the number of striae and their shape near the central area and the valve shape. A few species have an isolated punctum (c.f. *N. mutica*) and others have a silica thickening or septum-like structure near the apex. The striae in the species described below

---

PLATE 24

1 *Navicula lanceolata* x 1900 with striae detail x 4000, 2 *N. rhynchocephala* x 2500, 3 *N. tripunctata* x 1800, 4 *N. viridula* x 1500, 5 *N. peregrina* x 750, 6 *N. cuspidata* x 800, 7 *N. placenta* x 1200, 8 *N. mutica* x 1500, 9 *N. minuscula* x 4000.

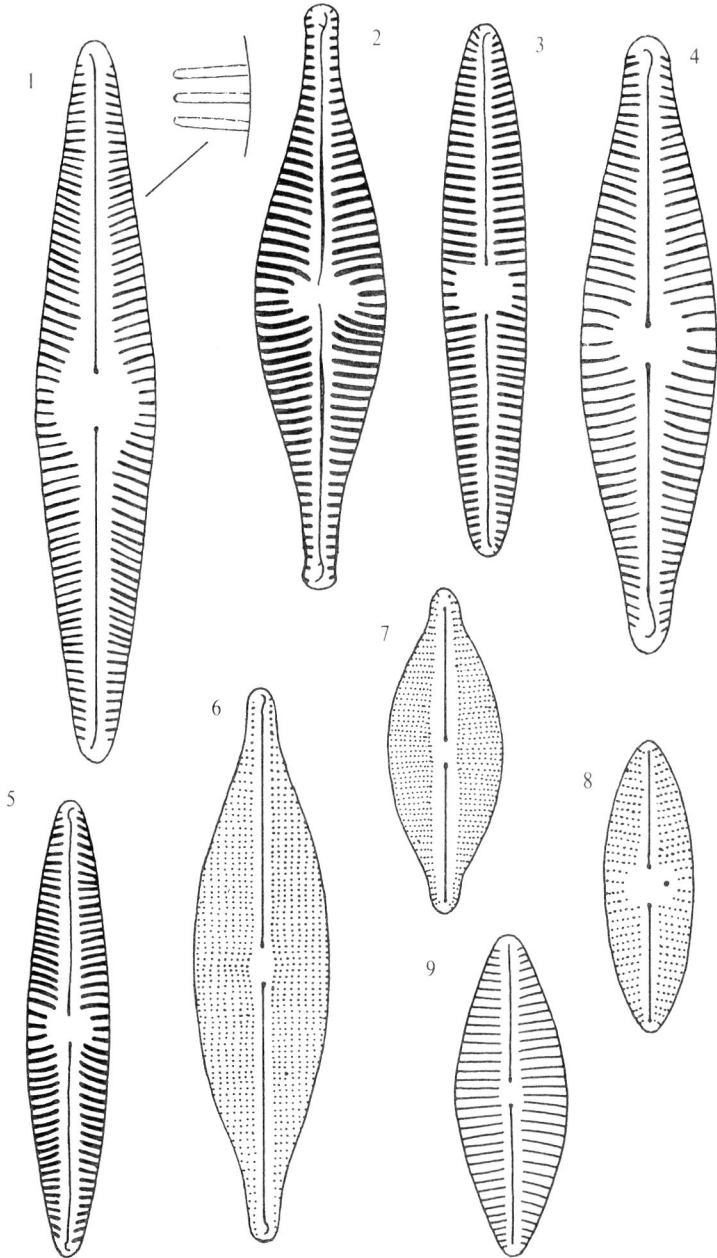

115

have a cross lineate structure (Plate 24 Fig. 1) and include some well known forms. They also share a roughly rounded central area and a narrow axial area.

1a   Apices capitate or rostrate, central area distinct, striae 8-10 ITM.   2

1b   Apices rounded or smoothly pointed.   3

2a   Valves lanceolate with capitate apices, 35-60 µm long, ltb 4-5. Striae parallel at apices.   *N. rhynchocephala*
(Plate 24 Fig. 2)

2b   Valves roughly linear, rostrate at apices, 40-80 µm long, ltb 4-5. Striae slightly convergent at apices.   *N. viridula*
(Plate 24 Fig. 4)

3a   Valves lanceolate, 40-150 µm long, ltb 4.5-6. Central area wide, striae 5-7 ITM.   *N. peregrina* (Plate 24 Fig. 5).

3b   Striae 10 or more ITM.   4

4a   Valves 27-50 µm, lanceolate, ltb 5. Striae radiate at the centre, 10-12 ITM with a large rounded central area.
N. lanceolata (Plate 24 Fig. 1).

4b   Valves linear near the central area, and slightly radiate, 11-12 ITM, almost parallel at the apices. Valves 30-60 µm, ltb 5.5-6.5, central area rectangular. *N. tripunctata*
(Plate 24 Fig. 3)

28. Gyrosigma

Valves broadly sigmoid with a narrow axial area and small central area. Striae finely punctate and parallel with a net-like appearance. Most species are benthic and they frequently exceed 100 µm in length. (Id: shape of valve apices, raphe structure, number of striae and their variation along the valve; 10).

G. attenuatum (Plate 27 Fig. 5) is a common species with the longitudinal rows of striae coarser (10-12 ITM) than the transverse rows (14-16 ITM). Valves 150-250 µm long, 7-8 ltb, apices symmetrically attenuated and smoothly rounded.

29. Cymbella

This is a large genus which contains many common species. The valves of *Cymbella* lack the bilateral symmetry of most other diatoms, giving them a pronounced dorsiventral appearance. The raphe either lies in a central position or close to the ventral margin (Plate 16, r). The axial and central areas vary between the species in both their size and position. The striae are frequently radiate on the dorsal margin but much reduced on the ventral side and vary from coarsely punctate to linear. Some species are solitary whilst others occur as filaments enclosed in mucilage tubes. (Id: valve ornamentation, valve shape, raphe position and shape; 140).

Key to the common species illustrated

1a Valves with one to several isolated punctae or stigmata in the central area.   2

1b Valves without isolated punctae or stigmata.   5

2a Valves with a single stigma on the dorsal side of the central area, strongly dorsiventral, 10-25 μm long, 2.5-3.5 ltb, ventral margin linear, apices slightly rostrate. Striae 28-35 ITM.   *C. minuta* (Plate 28 Fig. 2).

2b Valves with slightly gibbous ventral margin, with 1-several stigmata in the ventral side of the central area.   3

3a Valves 40-120 μm long, 2.5-3.5 ltb with 1-5 stigmata and punctate striae 18-20 ITM.   *C. cistula* (Plate 27 Fig. 1)

3b Valves 20-50 μm, slightly capitate, 3-4.5 ltb. Striae indistinctly punctate, 9-11 ITM.   4

4a Ventral striae shorter than dorsal striae, with one isolated stigma.   *C. affinis* (Plate 26 Fig. 5).

4b Striae about the same length on each valve, with two stigmata.   *C. turgidula* (Plate 26 Fig. 6)

5a Valves capitate.   6

5b Valve apices rounded or slightly rostrate.   8

6a   Raphe appearing as a single line with the apical ends bent towards the ventral margin. Valves 10-20 µm, 3.5-4.5 ltb, striae, 20-25 ITM.   *C. microcephala* (Plate 26 Fig. 7).

6b   Raphe appearing as a double line along part of its length, valves weakly dorsiventral, striae 12-18 ITM.   7

7a   Central area hardly wider than the axial area, raphe bent towards the ventral margin. Valves 25-35 µm, 6-7 ltb, striae 16-18 ITM.   *C. angustata* (Plate 26 Fig. 3)

7b   Central area wider than the axial area, valves 30-40 µm, 3.5-4.5 ltb, striae 12-14 ITM.   *C. naviculiformis* (Plate 27 Fig. 3)

8a   Striae fine, about 30 ITM with indefinite punctae. Valves weakly dorsiventral with indistinct central area, 16-30 µm long, 4.5-5.5 ltb.   *C. delicatula* (Plate 27 Fig. 2)

8b   Striae coarsely punctate, 12-20 ITM.   9

9a   Axial area narrow and curved throughout. Valves 100-210 µm long, 5.5-7 ltb. Raphe broad and bent towards dorsal margin at apices. Striae 9-10 ITM.   *C. lanceolata* (Plate 26 Fig. 4)

9b   Axial area broad and straight except at valve apices. Valves 40-80 µm, 3-4 ltb. Raphe ends bent towards the ventral margin. Striae 7-10 ITM.   *C. prostrata* (Plate 27 Fig. 4)

30. Amphora

The frustules of *Amphora* have a peculiar structure. The two valves are strongly arched and attached by their ventral margins so that the cells look like two *Cymbellas* stuck together. The raphe system of both valves appears on the same side so that the

---

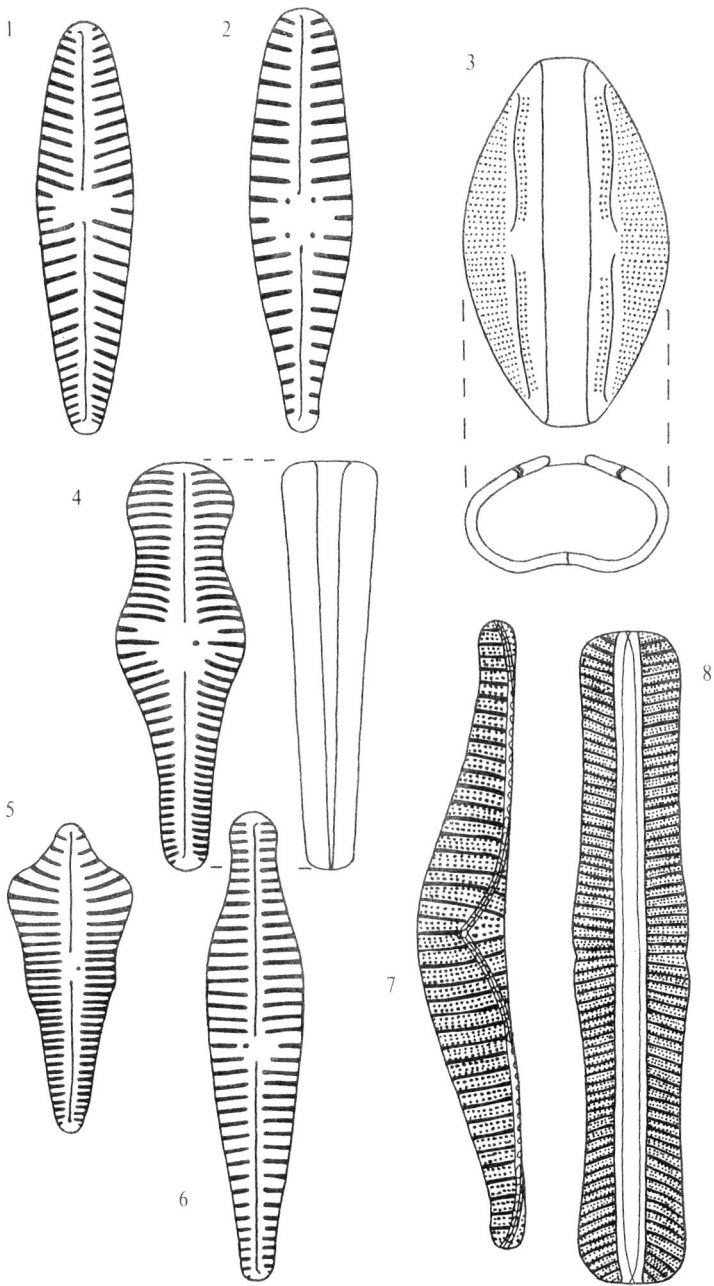

reverse side of the frustule, known as the *residuum* side has no raphe but just the remainder of the valve ornamentation. The ornamentation on the valves is difficult to see because of the curvature of the valves. (Id: valve and raphe shape, ornamentation; 10).

A. ovalis (Plate 25 Fig. 3) is a common species where the valves have round, non-capitate apices. The striae are punctate, 10-12 ITM, the central area extends to the ventral margin and the axial area is narrow. The frustules measure 30-90 µm long, 1.5-2 ltb. Frequent in hard water, benthic.

31. Gomphonema

The frustules of *Gomphonema* are somewhat club-shaped in valve view and wedge-shaped in girdle view. Many species are common epiphytes and they are often found attached to plants by mucilaginous stalks. In common with *Cymbella*, one group of species has isolated punctae present in the central area. (Id: valve shape, axial area and valve ornamentation; 50).

Key to the illustrated species

1a   One or more isolated stigmata present at the edge of the central area.   2

1b   Isolated stigmata absent. Valves boat-shaped and broadened towards one end, 15-40 µm long, 3.5-4.5 ltb. Striae radiate near the rectangular central area, 11-14 ITM.
    G. olivaceum (Plate 25 Fig. 1)

2a   Four or more isolated punctae in central area, two on either side. Valves club-shaped, widest just above the central area, 20-35 µm long, 3.5-4.5 ltb. Striae slightly radiate, 10-14 ITM.   *G. olivaceioides* (Plate 25 Fig. 2)

2b   Single stigma present on one side of the central area.   3

PLATE 26

1 *Didymosphenia geminata* Pen-y-Ghent, Yorks. x 900, 2 *Cymbella minuta* Uckfield, Sussex. x 2000, 3 *C. angustata* x 2900, 4 *C. lanceolata* Malham Tarn, Yorks. x 750, 5 *C. affinis* x 1700, 6 *C. turgidula* x 2000, 7 *C. microcephala* x 4000

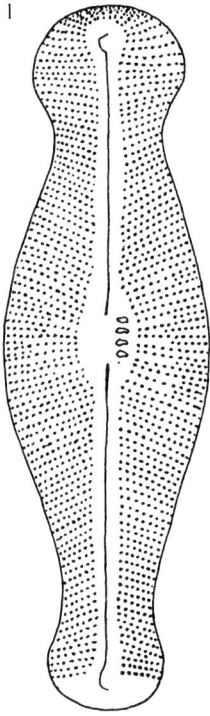

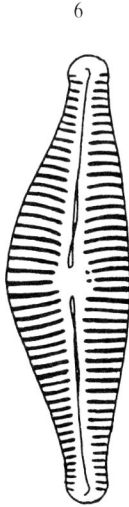

121

3a   One valve apex strongly wedge-shaped, the other narrowly rounded. Valves 30-90 µm, 2-3 ltb with indistinct central area. Striae radiate, 8-11 ITM.   *G. acuminatum* (Plate 25 Fig. 5)

3b   Valve apices otherwise.   4

4a   One apex much wider than the other and a distinct central area. Valves 25-65 µm, 2-3 ltb. Striae radiate, 10-12 ITM. *C. truncatum* (Plate 25 Fig. 4).

4b   Valve apices differing little in size, central area indistinct. Valves 20-30 µm, 4-5 ltb. Striae radiate, not surrounding valve apices, 11-13 ITM.   *G. angustatum* (Plate 25 Fig. 6).

32. Didymosphenia

The only species looks like a large *Gomphonema* but the ornamentation is much coarser and consists of punctate loculi interposed with costae. There is also a distinct unornamented area at the apex of the valve at the narrow end. *D. geminata* (Plate 26 Fig. 1) is a fine but variable diatom which sometimes occurs in large numbers in hard water streams, particularly in hilly districts. The valves are 100-150 µm long, 3-3.5 ltb with broadly capitate apices and one apex considerably wider than the other. The central area is ovate with 2-5 elongate pores at one side and the coarse striae are 8-10 ITM.

33. Rhopalodia

This is another genus with unusual valves which are normally seen only in girdle view, where the frustule appears rectangular with rounded ends and is usually centrally gibbous. In valve view the cells are sickle-or kidney-shaped with the raphe confined to the dorsal margin. The axial area is narrow or indistinct and the central area absent. The ornamentation consists of a series of transverse costae with two or more rows of intervening punctae which can be seen in both valve and girdle view. (Id: girdle shape, ornamentation; 6).

R. gibba (Plate 25 Fig. 8) is a frequent species with valves 80-300 µm long, 5-6 ltb in girdle view, which is rectangular, centrally gibbous with a slight waist. The costae are 6-8 ITM separated by two rows of punctae.

34. Epithemia

Valves dorsiventral with rounded, rostrate or capitate apices. Raphe gable-like, meeting the ventral margin near the apices and with an obscure, banded appearance. Transverse costae well developed with intervening fine striae, central and axial areas absent, girdle view rectangular. A small genus of attached diatoms. (Id: valve shape, costae shape, striae number between costae; 50).

E. turgida (Plate 25 Fig. 7) is a common species with valves 60-150 µm long, 6-7.5 ltb. The capices are rostrate and the ventral margin linear. The costae are 3-5 ITM with 2-3 rows of intervening punctae. In this species, the costae do not appear capitate when seen in girdle view.

35. Denticula

Valves lanceolate or elongate-ellipsoidal, often slightly asymmetrical. The species resemble *Diatoma* but a narrow, beaded raphe extends from one apex to the other and there is no axial or central area. The raphe in valve view is usually slightly curved and off-centre. Ornamentation consists of transverse costae with intervening rows of striae. (Id: raphe and costa shape, number of intervening striae; 50).

D. tenuis (Plate 22 Fig. 7) is a species often found among aquatic mosses. The valves are lanceolate, 8-60 µm long, 6-8 ltb with broad costae 5-7 ITM and fine striae 22-30 ITM.

36. Nitzschia

A large genus characterized by the peculiar raphe and valve structure. The valves are dorsiventral and often long and narrow. The raphe is displaced to one side and usually lies along the ventral margin where it appears to be crossed by a distinct row of punctae, called *carinal dots*. There is no central area and the axial area is narrow. The valves are traversed by fine striae which are sometimes unresolvable with the light microscope. A section through the valves is rhomboidal (Plate 27 Fig. 7) and the raphe is seen to be set into a keel so that both 'valve' and 'girdle' views can usually be seen at the same time. (Id: valve shape, shape and position of carinal dots, ornamentation; 200). Many of the species occur upon or among loose sediments in ponds, rivers and lakes. Others are found in the plankton of shallow lakes or among aquatic vegetation.

Key to illustrated species

1a   Valves with spine-like apices, cells 50-150 µm long, 18-25 ltb with unresolvable striae. Carinal dots 17-20 ITM, common.   *N. acicularis* (Plate 28 Fig. 3)

1b   Valve apices not spine-like   2

2a   Valves large and bilobate, 80-150 µm long, 5-7 ltb, carinal dots 5-7 ITM, striae 17-19 ITM.   *N. bilobata* (Plate 28 Fig. 2)

2b   Valves not bilobate   3

3a   Valves slightly sigmoid, 50-1000 µm long, 15-18 ltb with narrow attenuated apices. Carinal dots 7-12 ITM, striae 20-30 ITM, common.   *N. sigma* (Plate 27 Fig. 7)

3b   Valves not sigmoid   4

4a   Valves broadly rounded, 20-40 µm long, 4-5 ltb, carinal dots 10-14 ITM, striae 30-40 ITM.   *N. communis* (Plate 28 Fig. 4)

4b   Valves fusiform with slightly capitate apices, 2-70 µm long, 8-10 ltb. Carinal dots 10-15 ITM, striae 35-40 ITM, common.   *N. palea* (Plate 27 Fig. 6)

PLATE 27

1 *Cymbella cistula* x 750, 2 *C. delicatula* x 3500, 3 *C. naviculiformis* x 1800, 4 *C. prostrata* x 950, 5 *Gyrosigma attenuatum* valve x 600 and striae x 1500, 6 *Nitzschia palea* x 1600, 7 *N. sigma* valve x 250 and transverse section x 900

(All diatoms magn. reduced by 0.75)

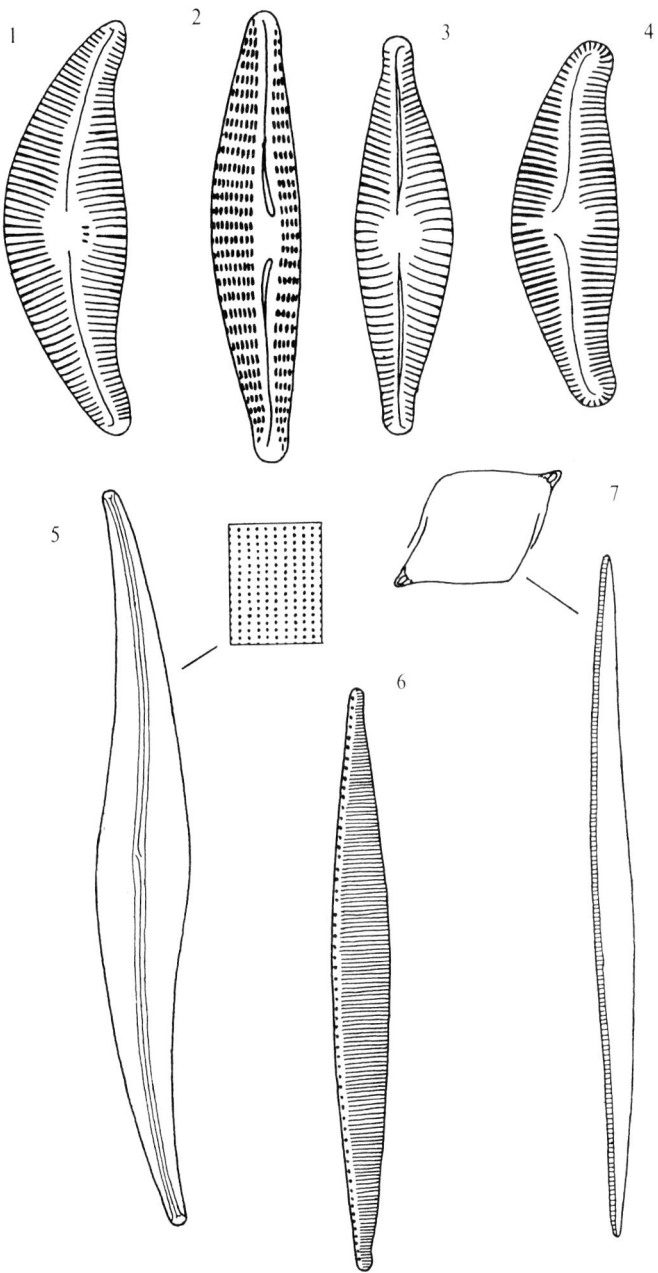

125

37. Surirella

Valves broadly ovate or club-shaped with the margins strongly costate. Fine transverse striae occur but they are not always resolvable and there is a keeled, marginal raphe. These large, solitary diatoms are easy to identify because of their coarse costae and the frilled appearance of the valve edges. In girdle view the valves are rectangular or slightly wedge-shaped with rounded corners. The species are usually benthic. (Id: shape in valve and girdle view, length and number of costae, number of striae; 100).

S. *biseriata* (Plate 22 Fig. 5) has lanceolate valves 80-350 μm long, ltb 2-3 with thick costae 1-2 ITM and unresolvable striae. It is rectangular in girdle view.

38. Cymatopleura

Valves elliptical and broadly waisted with a series of strong marginal ribs penetrating only slightly into the face of the valve. Raphe marginal, among the apices of the ribs. Girdle view narrow and slightly undulating. The undulations are also evident in the valve view as a series of transverse shadows. The striae are transverse and fine, not always resolvable. (Id: valve shape, ornamentation; 5).

C. *solea* is probably the best known species with broadly waisted valves and slightly pointed apices, 30-300 μm long, 4-6 ltb. The costae are 6-9 ITM. It is frequently found in the benthos of lakes (Plate 22 Fig. 6).

39. Campylodiscus

Frustules orbicular in valve view with strong marginal costae which converge towards the valve centre. The raphe is indistinct and set in the valve margin. These diatoms are solitary and extremely narrow in girdle view where they are often somewhat saddle-shaped. There are few species and they are not common. (Id: ornamentation, length and position of costae; 4).

C. *noricus* (Plate 17 Fig. 8) has strong, radiating costae which almost reach the valve centre. At the perimeter they are 2-3 ITM and appear forked. Very fine radiating striae are also present. Valves 60-150 μm wide.

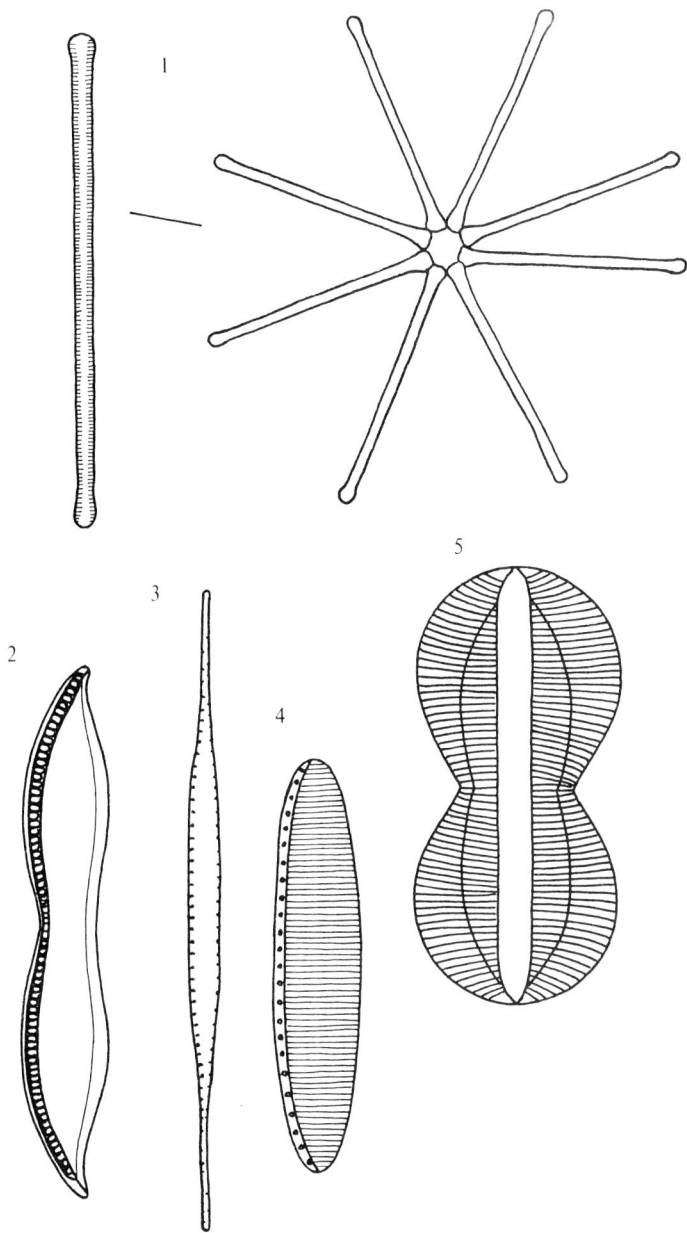

PLATE 28

1 *Asterionella formosa*, Windermere, colony x 300, cell x 675, 2 *Nitzschia bilobata*, x 800, 3 *N. acicularis* x 1000, 4 *N. communis*, x 2100, 5 *Entomoneis paludosa*, Romney, Kent x 1100

40. Entomoneis

The frustules of these species have distinctive keeled 'wings'
are normally only seen in girdle view where the marginal raphe is
not visible. The striae are usually finely punctate and the valves
are lightly silicified. (Id: wing structure, valve ornamentation;
10), (Syn. *Amphiprora*).

E. paludosa (Plate 28 Fig. 5) is the most frequent British
species with valves 35-130 μm long and an unchambered keel.
The striae are fine, 20-24 ITM. The species are all benthic and
normally occur in coastal districts.

THE CHLOROPHYCEAE OR GREEN ALGAE

This is the largest and most diverse class of freshwater algae containing forms ranging in size from bacteria (e.g. *Chorella minutissima*) to filaments over a metre in length (*Cladophora*). The main features of the class are the grass-green chloroplasts and the positive starch reaction although there are notable exceptions such as *Trentepohlia* where the green colour is masked by the production of excess β-carotene and the storage products are oils. The freshwater genera may be conveniently divided into nine orders and the distinguishing characters of these are shown in Table 3. This arrangement of classification broadly follows that of other authors although there is still much disagreement as to the number and size of the orders. A brief description of each order is given below.

The Volvocales are a large and well defined group of algae which are motile in their vegetative state. Most of the species possess two smooth flagella and swim in breast-stroke fashion, perhaps the best known example being *Chlamydomonas* whose biology has been extensively investigated. Even in this 'simple' example however, there is much variation in reproductive behaviour with a tendency towards oogamy in some of the species. The most structurally advanced form is *Volvox* which has a complex reproductive process involving an inversion stage.

The Tetrasporales is a closely related but much smaller order. Most of the species are colonial and each cell possesses one or more *pseudocilia*. These are extremely fine filaments resembling flagella but incapable of movement and their function remains obscure. They have been shown to differ in structure from flagella under the electron microscope. The cells in this order usually possess two contractile vacuoles near the base of the pseudocilia and the chloroplast is cup shaped. The life cycles of a few species have been investigated and found to be complex. An alternation of forms has been found in *Tetraspora gelatinosa* and includes an *Apiocystis* stage. Reproduction by biflagellate zoospores often occurs.

The Chlorococcales are a large and diverse order where most of the members are colonial. One feature is reproduction by autospores which occurs in at least two ways. In *Pediastrum* (Plate 33 Fig. 1) the cells produce bulbous vesicles into which a large number of zoospores are released. These then rearrange themselves within the vesicle to form a miniature colony and shed their flagella. Shortly afterwards the plates take on their characteristic form and are released through the breakage of the vesicle wall. This process is also found in *Hydrodictyon* where an entire net is assembled within the large vesicular cells. This method is found in only a few genera and in most forms the autospores are produced without an obvious zoospore stage and no vesicles are secreted. After the autospores are released, each cell enlarges until a full sized plant is produced. It is therefore common to find a whole range of sizes for any particular species within one collection of material. A few genera reproduce by zoospores as well as autospores and a small number normally reproduce only by simple fission (e.g. *Coccomyxa*). The chloroplasts are usually parietal, one per cell and with a single pyrenoid. The cells walls of some species are angular and thickened and sometimes tinted red or orange with accessory pigments. Resting cells are often observed.

In the Ulotrichales a simple filamentous or foliose form is seen in most genera with little differentiation between the basal, attachment cells and the erect system. All of the cells, with the exception of the basal cell are capable of subdividing to produce zoospores or gametes although a few genera are oogamous (e.g. *Sphaeroplea*). The zoospores usually possess four flagella and from 2-32 may be released from each cell, depending upon the cell size. Biflagellate isogametes are produced in a similar manner and some species form ' + ' and '−' gametes on separate plants.

The Chaetophorales is a large order of filamentous algae, some of which are much reduced in size and may consist of irregular packets of cells (e.g. *Desmococcus*). In a typical form, however, such as *Stigeoclonium*, the thalli consist of a basal, cushion-like growth attached to the substratum, which supports an erect system of much branched filaments. This differentiation into a basal and erect system is termed *heterotrichy* but its development varies greatly from one genus to the next. Another character shared by many species is the development of hairs from some of the cells, particularly the terminal ones, although in *Aphanochaete* they arise from the cells producing the greatly reduced prostrate system. There is much variation in the reproductive processes but most species produce quadriflagellate zoospores from unspecialized sporangia. Sexual reproduction ranges from isogamy to an advanced form of oogamy. Some members of the group exhibit a considerable division of labour and it has been suggested that the land plants arose from this stock. *Fritschiella tuberosa* is often cited to support this, where a rhizidial system supports a number of storage cells and an upright system which appears above the surface of tropical soils.

The Oedogoniales comprise another interesting order with an unique kind of cell division which results in the development of a series of annular rings near the cross walls of some of the cells. These result from the unequal division of the cells which requires one half of the cell to undergo an expansion, causing the rupture of the outer cell wall. The results of these frequent ruptures are a series of annular caps which are usually easy to see, even under low power. The sexual reproduction is also peculiar and complex. The male and female cells may be produced on either the same or separate plants and there are various ways by which fertilization is achieved (see p.189). Although the order has only three genera the number of species is large.

The next two orders possess large, multinucleate cells although the nuclei usually cannot be seen without special staining. In the Siphonocladales, cross walls occur in the freshwater species and the cells have a coarse feeling which is uncommon in algae. The cells are usually much longer than wide and contain a large net-like chloroplast with numerous pyrenoids. The species are all filamentous, often branched and attached by a basal holdfast. Asexual reproduction is by bi- or quadriflagellate zoospores and sexual reproduction is isogamous. In some species the ' + ' and '−' gametes are formed on different plants and an alternation of haploid and diploid generations has been recorded. The Dichotomosiphonales is represented by the sole genus *Dichotomosiphon* which has vesicular, well branched thalli without cross walls. It is closely related to the marine order Siphonales.

TABLE 3

Order	Section	Motile	Pseudocilia	Unicellular	Colonial	Cushion-forming	Foliar	Simple filaments	Branched filaments	Siphonaceous	Cell multinucleate
						Morphology and cytology					
Volvocales	A	✓		✓	✓						
Tetrasporales	B		✓	✓	✓	✓					
Chlorococcales	C			✓	✓						
Ulotrichales	D					✓	✓	✓			
Chaetophoralis	E					✓			✓		
Oedogoniales	F							✓	✓		
Siphonocladales	G							✓	✓		✓
Dichotomosiphonales	H								✓	✓	✓
Zygnematales Euconjugateae	I(1)							✓			
Desmidioideae	I(2)			✓	✓			✓			

Sexual reproduction			Special features	Page
Isogamous	Oogamous	Conjugation		
✓	✓		Chloroplast usually cup-shaped	
✓			Chloroplast cup-shaped	
✓	✓		Reproduction frequently by autospores, great range of cell and chloroplast shape.	
✓	✓		Chloroplast usually parietal and band-shaped	
✓	✓		Filaments often with hair-like ends. Some spp. orange-coloured	
	✓		Cells divide to form caps. Chloroplast parietal and net-like	
✓			Coarse with long cells. Chloroplast parietal and net-like	
	✓		*Dichotomosiphon* only genus. Filaments branched and tubular	
		✓	Chloroplasts helical, band-like or stellate, cells often much longer than wide.	
		✓	Cells often elaborate and waisted, chloroplasts mainly two, 1 in each semicell	

The Zygnematales is a large order whose members share one feature in common, reproduction by conjugation resulting in the formation of thick walled zygospores (Fig. 4c). The order is conveniently divided into two suborders. The Euconjugatae contain only filamentous forms showing simple cell division. The cells are often long and they are never constricted in the midregion. The best known example is *Spirogyra* which is easily recognized by its helical chloroplasts. In Britain, at least, the helices always appear to be left-handed. There are numerous species in both suborders and many, particularly in the Euconjugateae, are distinguished by their reproductive characters. Sexual reproduction is not often observed in temperate regions but it is sometimes possible to induce it in collected material by allowing it to dry out, altering the temperature or adding a small amount of salt.

The remaining suborder, the Desmidioideae contain the desmids, the majority of which are unicellular. These algae can be further subdivided on cell wall structure. The Placoderm desmids, to which most species belong, have a cell wall which is divided into two symmetrical *semicells* (see Plate 41 for details of morphology) connected by a narrow *isthmus* containing the nucleus. Each semicell usually contains one or two axile chloroplasts and these may be ribbed or notched making the cells fine microscopial objects. The cells of most Placoderm desmids are flattened to some extent so that they normally lie with their broad face and median constriction in *face view*. Some characters useful in identification are best seen in *vertical view* or *side view*. In the former, the median constriction cannot be seen (Plate 41 Fig. 12). The cell wall is often composed of two pieces although this is not always evident but the walls are nearly always covered with granules, teeth, spines or rows of pores. Cell division is complex and originates as an ingrowth of the median constriction as illustrated for *Cosmarium* (Plate 41 Fig. 9). In contrast, the Saccoderm desmids are far more simple in construction. They have capsule-shaped cells without a median constriction, no ornamentation and a simple form of cell division. A good example is *Mesotaenium* (Plate 42 Fig. 2). Some desmids are sufficiently large to be visible with the naked eye. The filamentous forms are readily distinguished from the Euconjugatae by the possession of a median constriction with a chloroplast in each semicell. The desmids are found in greatest abundance and diversity in nutrient-poor districts, particularly in montane regions where there is peaty soil. The lochs and bogs formed over the Lewisian rocks of NW Scotland are said to be particularly rich. Although it is often stated that desmids are uncommon in calcareous districts, this is not true of the British Carboniferous Limestone where they can sometimes be found in abundance.

Mattox and Stewart (1984) have proposed a new class-level system of classification for the Green algae based upon comparative cytology. They recognize five classes, namely the Micromonadophyceae, Charophyceae, Ulvophyceae, Pleurastrophyceae and the Chlorophyceae.

Section A Volvocales

1a	Unicellular	2
1b	Colonial, of 4 or more cells	12
2a	Cells with two flagella	3
2b	Cells with four flagella	11

3a Cells orange to deep red in colour, with the green chloroplast normally showing through. Envelope broad with radiating lines. Common in limestone pools. *Haematococcus*

| 3b | Cells otherwise, usually green. | 4 |

4a Cells enclosed within an outer envelope or theca which may be calcified and tinted brownish. 5

| 4b | Cells not enclosed within a theca. | 6 |

5a Theca spherical or spheroidal and minutely granular. *Coccomonas* 8

5b Theca composed of two lens-shaped valves, surface granular. *Phacotus* 9

| 6a | Cell wall smooth | 7 |
| 6b | Cell wall lobed | 10 |

7a	Cells with a broad, clear, outer envelope. *Sphaerellopsis* 3
7b	Cells without a broad envelope	8

7a   Cells with a broad, clear, outer envelope. *Sphaerellopsis* 3

7b   Cells without a broad envelope   8

8a   Cells fusiform with pointed apices, up to 180 μm long.
     Chlorogonium 5

8b   Cells spherical, spheroidal, cylindrical or shortly fusiform
     with rounded apices.   9

9a   Chloroplast without a pyrenoid.   *Chloromonas* 2

9b   Chloroplast with 1-several pyrenoids   *Chlamydomonas* 1

10a Cells with 4-8 rounded lobes seen in side and vertical view.
     Lobomonas 6

10b Cells with two broad, wing-like lobes in side view and four
     lobes in vertical view, (rare).   *Brachiomonas* 7

11a Cells spherical, spheroidal or shortly cylindrical with 4
     flagella arising from a rounded apex.   *Carteria* 10

11b Cells lobate with the flagella arising from a broadly
     flattened and slightly depressed apex.   *Pyramimonas* 11

12a Cells 4-16 in a regular, flat plate   *Gonium* 12

12b Cells in a three dimensional colony   13

13a Colonies consisting of hollow spheres composed of large
     numbers of cells, usually containing several spherical
     daughter colonies.   *Volvox* 16

13b Colonies with < 200 cells, no daughter colonies.   14

14a Cells closely packed and angular, not separated by
     mucilage, 8-32/colony.   *Pandorina* 13

14b Cells regularly dispersed in spherical colonies, each cell
     separated by mucilage.   15

15a Cells spherical or spheroidal, and 8-128/colony. *Eudorina*
     14

15b Cells elongate with spike-like projections, usually in tiers of
     4-8 in small colonies, uncommon.   *Stephanosphaera* 15

1. Chlamydomonas

Although a huge number of species have been described on morphological grounds, it is doubtful whether these all represent genetically distinct forms. Species of *Chlamydomonas* are often abundant in small pools and ditches particularly among aquatic plants. The cells contain a large, frequently cup-shaped chloroplast within which a reddish stigma is usually apparent and between the two flagella, the cell wall is frequently drawn out into a small *papilla*. Most species belong to the section *Euchlamydomonas* but representatives from the sections *Chlamyedella*, *Chlorogoniella* and *Bicocca* are also frequently encountered. A key to the main sections with some common representative species is given below. For identification, particular attention should be paid to the structure of the chloroplast and position of the pyrenoids. Intermediate forms which defy identification are frequently encountered. Mention should also be made here of a related genus, *Dunaliella* which may form reddish blooms in ponds near the sea. The cells differ from *Chlamydomonas* in lacking a cell wall and showing metabolic movements. (Id: cell shape, chloroplast structure, presence and size of stigma and papilla, position of contractile vacuoles; z (4-8), I, O; 500). In common with other Volvocales, non-motile *Palmella* stages often occur.

Key to sections and representative species

1a	Chloroplast with a single pyrenoid.	2
1b	Chloroplast with 2-several pyrenoids.	6
2a	Chloroplast a simple cup	3
2b	Chloroplast otherwise	4
3a	Pyrenoid in a basal position.	*Euchlamydomonas.*

Most of the species encountered will belong here. *C. debaryana* (Plate 29 Fig. 4) has spheroidal cells 12-17 µm long, 1.2-1.6 ltb with a distinct, rounded papilla, roundish stigma and spherical pyrenoid. The rim of the chloroplast cup comes close to the cell apex. *C. reinhardtii* (Plate 29 Fig. 1) differs in lacking a papilla and having spherical cells

15-25 µm d. although the pyrenoid and chloroplast is similar. Less frequently reported is *C. snowiae* with elongate-spheroidal cells, 15-20 µm long, 2 ltb and finely striated chloroplast. An acute papilla is also present (Plate 29 Fig. 2).

3b   Pyrenoid lateral.   *Chlamydella.*
      C. gloeogama has spheroidal cells 12-20 µm long, 1.5-1.9 ltb with a cup-shaped, smooth chloroplast and a small conical papilla (Plate 29 Fig. 5)

4a   Chloroplast against the side wall with a lateral pyrenoid.
                                                          Chlorogoniella.
      C. acidophila (Plate 29 Fig. 3) can sometimes be collected among *Sphagnum* with other species. The cells are spheroidal, 8-11 µm long, 1.6-2 ltb with a large lateral chloroplast incompletely encircling the wall. There is no papilla.

4b   Chloroplast H-shaped with a pyrenoid in the bar of the H.
                                                                       5

5a   Nucleus apical.   *Pseudagloe.*
      C. stellata has an H-shaped chloroplast which is broken up into a small number of polygonal fragments. Cells spheroidal, 18-22 µm long, 1.4-1.6 ltb with a small rounded papilla (Plate 29 Fig. 7).

5b   Nucleus basal.   *Agloe.*
      C. regularis is a typical species with a large central pyrenoid in the bar of the H. Cells spheroidal, 20-30 µm long, 1.4-1.7 ltb with a smooth chloroplast and conical papilla (Plate 29 Fig. 6).

PLATE 29

1 *Chlamydomonas reinhardtii*, Ashdown Forest, Sussex, 2 *C. snowiae*, 3 *C. acidophila*, 4 *C. debaryana*, Newhaven, Sussex, 5 *C. gloeogama*, Newhaven, Sussex., 6 *C. regularis*, 7 *C. stellata*, 8 *C. platyrhyncha*, Tonbridge, Kent., 9 *C. gallica*, Malham, Yorks., 10 *C. pertusa*, 11 *C. sphagnicola*, 12 *Chloromonas inversa*, Cuckmere, Sussex, 13 *Sphaerellopsis gelatinosa*, Goudhurst, Kent, 14 *Haematococcus pluvialis*, Malham, Yorks.

All x 1700 (.75)

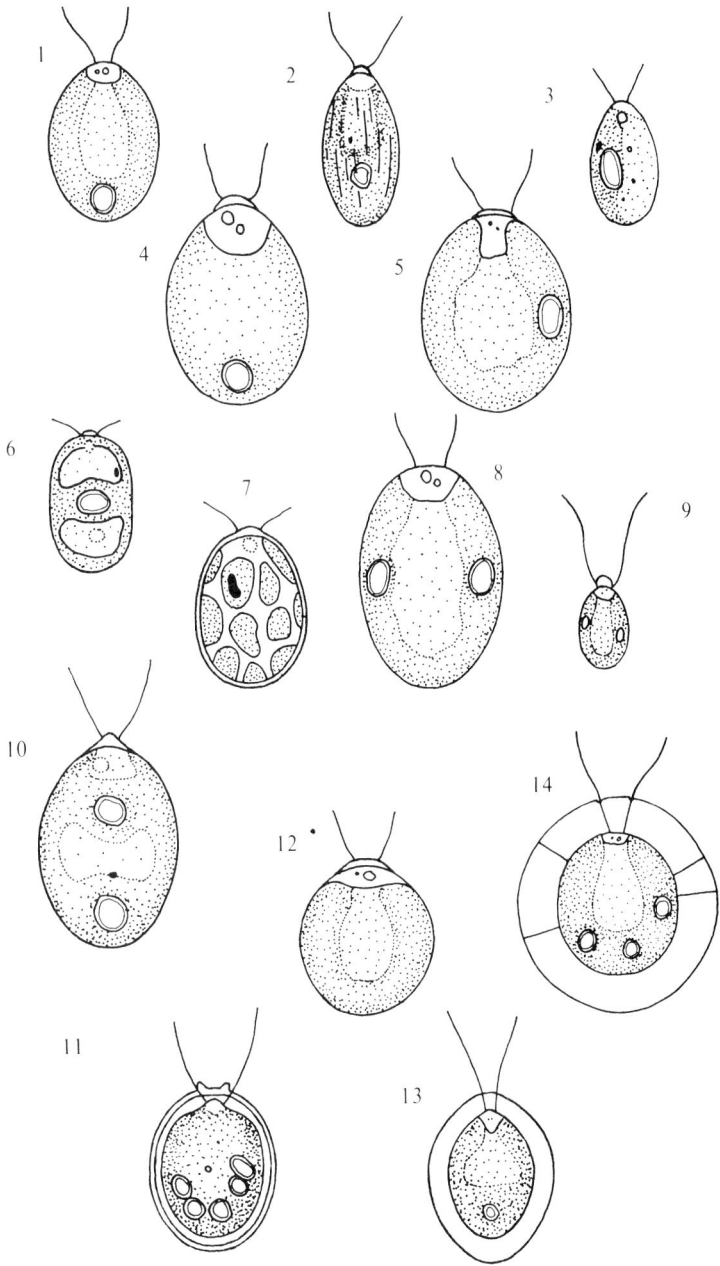

6b Cells with 3-many pyrenoids. *Pleiochloris.*

C. sphagnicola has a large, smooth, cup-shaped chloroplast containing 4-6 pyrenoids. The cells are spheroidal, 25-35 µm long, 1.2-1.5 ltb with a truncated and divided papilla (Plate 29 Fig. 11).

7a Chloroplast cup-shaped, with two lateral pyrenoids.
Bicocca.

A small but distinctive section containing several frequent forms. *C. platyrhyncha* has spheroidal cells with a small rounded papilla, 18-25 µm long, 1.3-1.6 ltb (Plate 29 Fig. 8). *C. gallica* is a species with a prominent, hemispherical papilla and rounded cylindrical cells, 25-32 µm long, 1.7-2 ltb (Plate 29 Fig. 9).

7b Chloroplast H-shaped with a double bar and a pyrenoid in each bar. *Amphichloris.*

Here the chloroplast structure is quite complex and variable.

C. pertusa (Plate 29 Fig. 10) is fairly well known with a smooth, unbroken chloroplast and a central rectangular clear area. Cells spheroidal or slightly pointed, 15-25 µm long, 1.5-1.9 ltb without a papilla.

2. Chloromonas

The cells are similar to *Chlamydomonas* but lack pyrenoids. Most species have a smooth cup-shaped chloroplast but in a few it is finely striated. The cells are spherical or spheroidal, with or without a papilla. (Id: chloroplast structure, cell shape, papilla; 30). Some forms are common in small ponds and ditches.

C. inversa (Plate 29 Fig. 12) is a species with spherical cells 14-19 µm wide and a prominent, rounded papilla. The chloroplast is smooth and cup-shaped.

3. Sphaerellopsis

Cells similar to *Chlamydomonas* but with a broad, hyaline envelope also seen in *Haematococcus*. (Id: chloroplast shape and structure, stigma size, papilla, cell shape; 15). The species are sometimes frequent in small water bodies.

S. gelatinosa (Plate 29 Fig. 13) has a smooth, cup-shaped chloroplast with a basal pyrenoid. The cells are somewhat egg-shaped, 15-20 µm long, 1.2-1.6 ltb, and there is no papilla.

4. Haematococcus

The cells of these species appear orange or reddish due to the presence of a carotenoid pigment, astaxanthin. The cells are most often found in an encysted state in the bottoms of calcareous hollows which dry out during the summer. The chloroplast is large and cup-shaped and a broad cell envelope is present. (Id: chloroplast shape, number of pyrenoids; 4).

H. pluvialis (Plate 29 Fig. 14) is a common British species with a cup-shaped chloroplast and 3-4 pyrenoids. The cells are spherical or spheroidal, 9-35(70) µm long, 1.1-1.4 ltb. Other species with just two pyrenoids are far less common.

5. Chlorogonium

Cells elongate and spindle-shaped with a long chloroplast with or without pyrenoids. The contractile vacuoles are not apical as they are in *Chlamydomonas*. The species sometimes occur in large numbers in small water bodies where they swim with undulating movements. (Id: number of pyrenoids and contractile vacuoles, cell and chloroplast shape; 30). The species range from 25-100 µm in length, 5-15(25) ltb.

C. elongatum (Plate 30 Fig. 1) is a frequent form with cells 30-60 µm long, 8-15 ltb. There are two pyrenoids and contractile vacuoles within the cells.

6. Lobomonas

Cells are of the *Chlamydomonas* type but with a regularly warty cell wall but the species are not well defined. *Lobomonas* is sometimes found in ponds and ditches, especially near the coast. (Id: cell wall shape, papilla, number of contractile vacuoles; 8). The cells measure 3-25(40) µm in long.

L. ampla (Plate 30 Fig. 2) has small, rounded and evenly

distributed warts, two apical contractile vacuoles and a smooth, cup-shaped chloroplast. The cells measure 12-22(40) µm long, 1.2-1.4 ltb. and no papilla.

7. Brachiomonas

The cells of these algae have pronounced, angular wings and there are few species, none of which is common. The chloroplast is parietal but often irregular in shape. (Id: development of wings, number of pyrenoids, papilla; 5).

B. submarina (Plate 30 Fig. 3) is the most frequently reported species with well developed wings and cells 15-25 µm long, 0.9-1.2 ltb. There is a single pyrenoid and small truncated papilla.

8. Coccomonas

Cells spherical to spheroidal, narrowed towards the apex and enclosed by a spherical, or angular calcified theca which may be brownish in colour. The species are normally found in hard water lakes, sometimes among vegetation but they are not common. (Id: shape of theca; 10).

A species of widespread occurrence is *C. orbicularis* (Plate 31 Fig. 4) with a spherical theca 20-50 µm wide with short longitudinal deposits of calcium carbonate apparent on the surface.

9. Phacotus

Cells approximately spherical with a cup-shaped chloroplast partly obscured by the heavily calcified lenticular theca. This genus is hard to mistake for any other and the species are

---

PLATE 30

1 *Chlorogonium elongatum*, Pevensey, Sussex, x 1600, 2 *Lobomonas ampla*, Pevensey, Sussex, x 1600, 3 *Brachiomonas submarina*, x 1500, 4 *Coccomonas orbicularis*, x 750, 5 *Phacotus lenticularis*, R. Coquet, Northumbria, x 1600, 6 *Carteria globosa*, x 1300, 7 *Pyramimonas tetrarhynchus*, Pevensey, Sussex, x 1600, 8 *Gonium pectorale*, Tunbridge Wells, Kent, x 750, 9 *Pandorina morum*, Speldhurst, Kent, x 750, 10 *Eudorina elegans*, Wray Mire Tarn, Cumbria, colony x 160 cell x 450.

(.75)

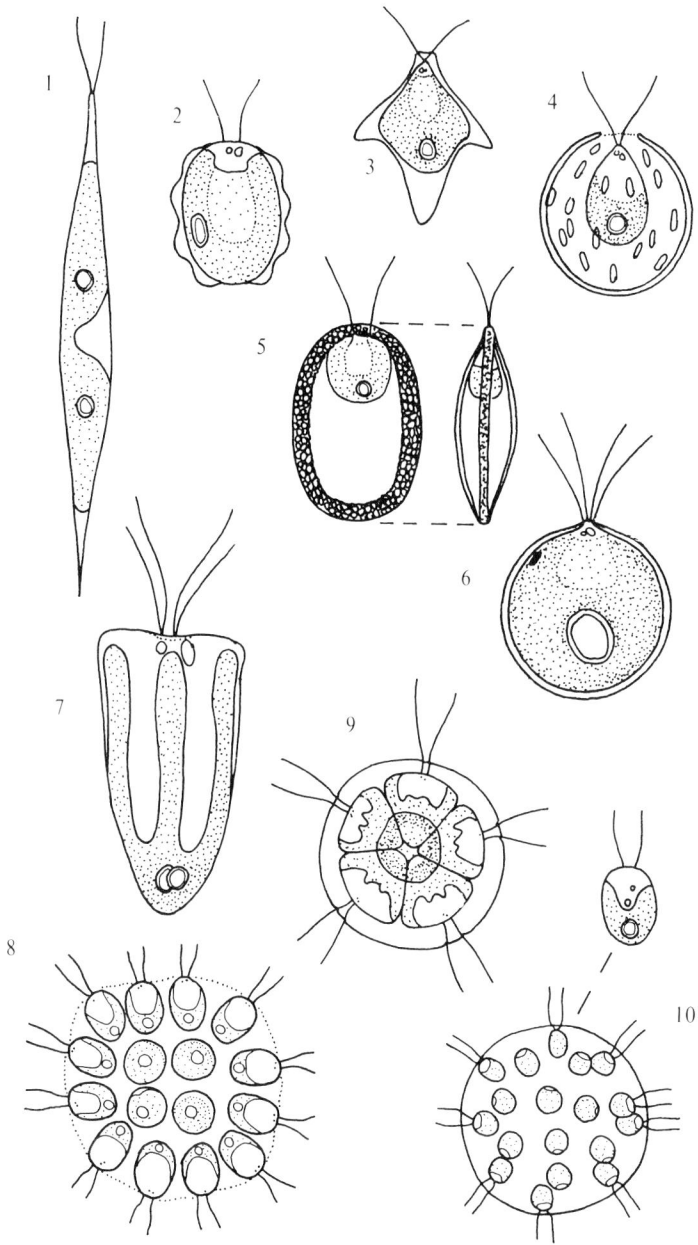

frequently found in hard water lakes or ditches. (Id: shape and structure of theca; 10).

P. lenticularis (Plate 30 Fig. 5) is a common British species with a coarsely granular, lens-shaped theca 12-30 µm in diameter. The chloroplast has a single basal pyrenoid.

10. Carteria

A genus closely allied to *Chlamydomonas* but differing in the presence of four flagella rather than two. The species occur in similar habitats to *Chlamydomonas* but they are less common. Species without a pyrenoid have been referred to the genus *Tetramastix*. The cells range from (4)6-30(70) µm in length, 1-4 ltb). (Id: chloroplast shape and structure, number of pyrenoids, stigma, papilla, cell shape; 60).

C. globosa (Plate 30 Fig. 6) is a species with a smooth, cup-shaped chloroplast and a basal pyrenoid. The cells are spherical, 18-28 µm wide and without a papilla.

11. Pyramimonas

The cells of these species are naked and covered in minute scales which can only be seen under the electron microscope. The chloroplast is parietal and either smooth or net-like and there is usually a basal pyrenoid. These algae are sometimes found in large numbers in cold water during the winter in ponds and ditches. (Id: cell shape, chloroplast shape; 8).

P. tetrarhynchus is a well defined species with four longitudinal channels and a broadly truncated apex which is gradually narrowed towards the rounded posterior end. The cells are 20-30 µm long, 1.8-2.5 ltb (Plate 30.Fig. 7).

This genus, together with a number of others which are mostly marine have been assigned to the class Prasinophyceae which are distinguished by their scaly walls and flagellar insertion.

12. Gonium

Colonies composed of 4-32 spherical or spheroidal cells arranged as a flat plate with the flagella directed out towards the margin of the plate. The species are sometimes abundant in small ponds and ditches.

G. pectorale (Plate 30 Fig. 8) is the commonest species with 16 cells arranged in a square with truncated corners. The cells are spherical and 7-12 µm wide. *G. sociale* has just four spheroidal cells 10-20 µm which are connected to each other by small growths of mucilage. An uncommon species is *G. formosum* which has 16 cells separated by broad areas of mucilage having a porose structure.

13. Pandorina

Colonies spherical, composed of 8-32 closely packed cells surrounded by a clear zone of mucilage from which the flagella protrude. The cell apices are broadly flattened and the chloroplast is cup-shaped and either smooth or striated. The colonies range up to 250 µm wide with cells 10-20 µm long. The species are usually found in small numbers with other Volvocales. (Id: chloroplast structure, number of pyrenoids; 3).

P. morum (Plate 30 Fig. 9) is the commonest with a finely divided or smooth chloroplast and usually a basal pyrenoid. The cells measure 12-17 µm long, 1-1.4 ltb.

14. Eudorina

Colonies composed of 16-64 cells arranged more or less regularly near the surface of a spherical mucilaginous investment. Cells usually spherical and often connected to each other by delicate strands which can be revealed by staining. The chloroplast is cup-shaped with 1-several pyrenoids. The species are sometimes seen in the plankton of ponds and lakes, particularly in hard water. (Id: colony shape, chloroplast structure number of pyrenoids; Aut, 0; 8).

E. elegans appears to be the most widely reported species. The colonies measure 60-200 µm wide and are composed of equal-sized spherical cells 16-24 µm in diameter. The chloroplast is smooth and cup-shaped with a basal pyrenoid (Plate 30 Fig. 10). *E. charkowiensis* is similar but has compressed cells with longitudinally striated chloroplasts.

15. Stephanosphaera

Colonies spherical or spheroidal containing a tier of 2-16 cells with the flagella directed outwards. The cells are elongate bodies with many spike-like protrusions of the wall giving them a prickly appearance.

The only species, *S. pluvialis* has cells 6-12 µm long, ltb 2.5-4.5 with an irregular parietal chloroplast and O-several pyrenoids, colonies 25-60 µm wide. It is occasionally found in limestone lakes and pools, sometimes associated with *Haematococcus* (Plate 31 Fig. 3).

16. Volvox

Colonies large and spherical with the cells regularly dispersed at the periphery. The cells number 250 to many thousand. Daughter colonies which develop from special cells are usually present within the hollow sphere. *Volvox* colonies measure 0.5-3 mm in diameter so that they can often be observed directly in the water but the cells are small and the chloroplast structure is often difficult to see. Sexual reproduction is often evident late in the season. The male cells or *sperm packets* are minute spindle-shaped bodies which proceed in small packets from enlarged vegetative cells (Plate 31 Fig. 1). The female cells are spherical and have thickened walls. After fertilization, the eggs often become orange in colour and may develop a spiny wall. Several species occur as male and female colonies which may differ in size. (Id: cell number per colony, chloroplast structure, presence of protoplasmic connections between the cells, sexes separate or on same plant; 20). Identification is difficult as there appears to be considerable variation within the species. Three forms are found in Europe. Cell numbers can be estimated using Equation 1 (appendix, p225).

PLATE 31

1 *Volvox aureus*, Malham Tarn, Yorks. Colony with four daughter cells x 100, cells showing protoplasmic strands x 900, cell in side view x 900, sp., sperm packet x 1500, 2 *V. tertius*, surface view of colony x 900, cell in side view x 900, 3 *Stephanosphaera pluvialis*, Malham Tarn, Yorks, x 1600, 4 *Paulschulzia pseudovolvox*, Llyn Padarn, Gwynedd. x 700, 5 *Tetraspora gelatinosa*, R. Coquet, Northumbria, colony x 1, group of cells x 1000, cell with pseudocilia x 1000, 6 *Chaetopeltis orbicularis* x 225.

(.75)

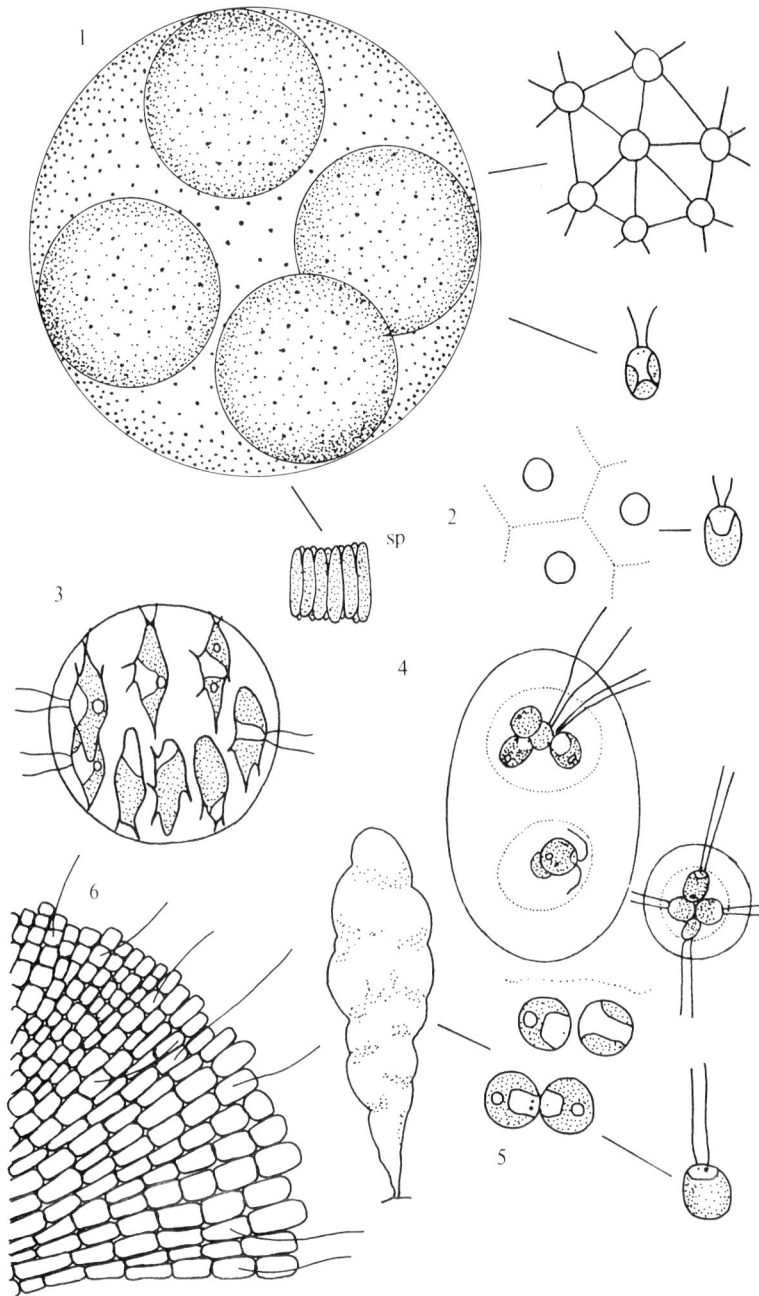

1

2

sp

3

4

6

5

147

1a   Cells interconnected by fine protoplasmic strands (stain with methylene blue).   2

1b   Protoplasmic strands absent but cells often separated from each other by a polygonal network of lines. Cells 4-8 μm wide with a cup-shaped chloroplast. Colonies 250-600 μm wide, containing 500-2000 cells. Sexes on different plants.   *V. tertius* (Plate 31 Fig. 2)

2a   Cells rounded, 4-6 μm wide in colonies of 500-3200 cells. Colonies 200-600 μm wide, males and females on separate plants.   *V. aureus* (Plate 31 Fig. 1)

2b   Cells somewhat irregular and stellate, 2-4 μm wide, colonies of 1500-25000 cells. Colonies 500-1000 μm wide. Cells separated from each other by a polygonal network of fine lines, sexes on the same plant.   *V. globator*

Section B. Tetrasporales

1a   Planktonic, consisting of small colonies of cells.   *Paulschulzia* 17

1b   Colonies large, growing on rocks and plants in shallow water.   2

2a   Plants consisting of a flat disc of cells closely attached to aquatic plants, many cells with 1-2 erect pseudocilia.   *Chaetopeltis* 20

2b   Colonies gelatinous, not disc-like.   3

3a   Colonies up to 1 mm long, pear-shaped with pseudocilia radiating outwards.   *Apiocystis* 18

3b   Colonies large and irregular, to 10 cm, loosely attached gelatinous masses with pseudocilia generally covered by mucilage.   *Tetraspora* 19

17. Paulschulzia

The colonies consist of 4-200 spherical cells normally enclosed by broad and faintly lamellate mucilaginous envelopes up to 200 μm wide. The chloroplast is parietal and cup-shaped with a pyrenoid. Two contractile vacuoles are present and each cell possesses two pseudocilia.

P. pseudovolvox (Plate 31 Fig. 4) has cells 6-14 μm wide with long pseudocilia and is sometimes abundant in the plankton of lakes. *P. elegans* differs in having the pseudocilia short and enclosed in the mucilage envelope.

18. Apiocystis

Colonies pear-shaped and attached by a narrow base to aquatic plants. The small spherical cells each possess two pseudocilia and they are often seen scattered in groups of four over the colony surface.

A. brauniana (Plate 41 Fig. 2) has cells 6-12 μm wide with a cup shaped chloroplast and basal pyrenoid. It is a frequent epiphyte of lowland ponds and lakes with colonies up to 1 mm long.

19. Tetraspora

Colonies large and irregular, sometimes tubular or cylindrical and resembling lumps of green jelly. The cells are distributed in groups of 2 or 4 throughout the colony and often number over a million per plant. Each cell usually has two fine and approximately parallel pseudocilia, together with a cup-shaped chloroplast and basal pyrenoid. The cells are confined mostly to the edge of the colony. (Id: colony shape and structure; z, I; 6).

T. gelatinosa is the most common form with irregular gelatinous thalli which may be lobed at the margins (Plate 31 Fig. 5). The cells are spherical and 6-12 μm wide. The species are widely distributed in shallow, often slow-flowing water. *Schizochlamys delicatula* is rather similar but distinguished by the presence of broken remains of old cell walls dispersed among the cells. The pseudocilia are extremely fine and often better seen after preservation in formalin.

20. Chaetopeltis

Thalli forming thin, orbicular patches up to 1 mm diameter on aquatic plants. The cells are angular and contain a cup-shaped chloroplast and two contractile vacuoles. Some cells possess one or two fine, erect pseudocilia.

C. orbicularis is the best known species with cells 15-20 µm wide (Plate 31 Fig. 6). *C. americana* is similar but the cells are smaller, 8-12 µm wide. The species should be compared with *Coleochaete* which they superficially resemble.

Section C. Chlorococcales

1a Colonies net-like, visible to the naked eye, composed of large cells with inflated ends arranged in an hexagonal framework.   *Hydrodictyon* 23

1b Colonies not net-like, cells microscopic.   2

2a Colonies of 5 or more cells arranged in a flat disc, the cells at the disc margin with 1-4 stumpy teeth or short spines.   *Pediastrum* 21

2b Colonies not disc-like with marginal teeth.   3

3a Cells arranged in short linear groups attached by their long faces. Outer cells often with two or more sharp spines, mucilage absent.   *Scenedesmus* 41

3b Cells not in linear groups   4

4a Cells radiating from a central point to which each cell is attached by a short stalk. Cells often angular and with 1-4 radiating spines (rare).   *Sorastrum* 22

4b Cells not in radiating groups attached by a narrow stalk. 5

5a Cells living at the water-air surface (neustonic). Cells solitary, spheroidal and sometimes with a thickened cap-like structure.   *Nautococcus* 54

5b Cells not neustonic   6

6a Cell walls with fine spines, ridges or other thickenings. 7

6b Cell walls smooth, rarely unevenly thickened, if cells tapered then not drawn out into colourless spines.   15

7a   Cells with long, narrow spines. Small planktonic algae. 8

7b   Cells with angular or striated thickenings.   11

8a   Cells fusiform, often slightly bent, with the apices drawn out into spines.   *Schroederia* 38

8b   Cells spherical or spheroidal.   9

9a   Unicellular   10

9b   Cells in small colonies, usually in sub-groups of four, each cell with 2-7 spines.   *Micractinium* 26

10a Cells with 2-8 regularly arranged spines.   *Chodatella* 24

10b Cells with numerous, irregularly arranged, radiating spines.   *Franceia* 25

11a Cells egg-shaped with two polar thickenings and several discoid chloroplasts.   *Oocystis* 31

11b Cells otherwise, with a single chloroplast.   12

12a Unicellular   13

12b Colonial, consisting of a regularly packed group of spherical cells often joined by short lateral processes. Colony spherical, often with regular marginal thickenings, mucilage absent.   *Coelastrum* 44

13a Cells angular and polyhedral, usually square or triangular in shape.   *Tetraedron* 39

13b Cells spherical or spheroidal   14

14a Cells spherical with thick walls which are covered with short teeth or regular thickenings, normally green.   *Trochiscia* 34

14b Cells elongate-spheroidal with a series of longitudinal ridges, green or red in colour, rare.   *Scottiella* 35

15a Unicellular, epiphytic algae attached by a thin stalk.   *Characium* 37

15b Algae not attached by a thin stalk   16

16a	Cells elongate, >2.5 ltb	17
16b	Cells short, <2.5 ltb	21

17a Cells in parallel bundles of four, long and narrow. *Quadrigula* 30

17b Cells not in parallel bundles of four. 18

18a Cells in mucilaginous colonies, with one end of the cell truncated, developing in pairs united by their truncated ends, contractile vacuoles maybe present.
Elakatothrix 46

18b Cells not in such pairs. 19

19a Cells fusiform, forming stellate colonies. *Actinastrum* 42

19b Cells not in stellate colonies, fusiform or sickle-shaped. 20

20a Cells solitary, large, with a row of pyrenoids in the long chloroplast. *Closteriopsis* 28

20b Cells solitary or in irregular bundles, sickle-shaped or fusiform with a single chloroplast. *Ankistrodesmus* 27

21a Cells large and living within other aquatic plants. *Chlorochytrium* 36

21b Cells not living within other plants. 22

22a Cells colonial, often in sub-groups of four, with the remains of old mother cell walls often present. Mostly planktonic. 23

22b Cells unicellular or if colonial then not normally in groups of four. 26

23a Cells densely or closely packed together in groups of 4 25

23b Cells in loose groups of four 24

24a Cells in small mucilaginous colonies, normally associated with mother cell wall remains. *Dictyosphaerium* 52

24b Cells spheroidal, with several discoidal chloroplasts per cell, cell wall remains normally absent. *Oocystis* 31

25a	Cells forming small, quadrate colonies without a wide mucilage envelope.	*Crucigenia* 43
25b	Cells packed in a tetrahedral arrangement and enclosed by a wide mucilaginous envelope. Mother cell wall remains sometimes present.	*Radiococcus* 53
26a	Planktonic or among submerged plants	27
26b	Terrestrial or growing in a thin layer on wet rocks. (cf. *Desmococcus* and *Apatococcus* which will key out here but belong to the Chaetophorales)	36
27a	Cells spherical or spheroidal	28
27b	Cells angular or sickle-shaped	35
28a	Cells in colonies	29
28b	Cells solitary	34
29a	Colonies composed of densely packed botryoidal masses of cells, 8-many per colony, (planktonic).	30
29b	Colonies not in botryoidal masses	32
30a	Colonies small and spherical consisting of 8-64 cells regularly arranged and closely packed.	*Coelastrum* 44
30b	Colonies of dense masses of cells or of small colonies held together in mucilage.	31
31a	Cells in dense masses, often dark green-brown or reddish in colour, held together in an oily mucilage.	*Botryococcus* 51
31b	Cells spherical, in small dense clumps held together in a wide mucilaginous envelope.	*Sphaerocystis* 48
32a	Cells spherical or hemispherical and often in pairs, with their flat faces opposite each other enclosed by a wide, sometimes lamellate mucilage envelope, with one chloroplast per cell.	33
32b	Cells spheroidal and usually enclosed within a mother cell wall, with 1-several discoidal chloroplasts per cell.	*Oocystis* 31

33a   Cells dispersed in pairs in a wide mucilage envelope, chloroplast parietal.   *Pseudosphaerocystis* 49

33b   Chloroplast axile and stellate, cells 1-4 enclosed in a firm lamellate envelope.   *Asterococcus* 50

34a   Cells large (§100 µm) and spherical containing numerous discoidal chloroplasts, frequent among wet *Sphagnum.*   *Eremosphaera* 33

34b   Cells small with a single parietal chloroplast, reproducing by autospores.   *Chlorella* 32

35a   Cells sickle-shaped, or lunate isolated or in small colonies.   *Kirchneriella* 29

35b   Cells angular, usually appearing square or triangular.   *Tetraedron* 39

36a   Cells often in large numbers, in mucilaginous colonies.   37

36b   Cells isolated or in dense, non-mucilaginous colonies.   38

37a   Cells spheroidal with a parietal chloroplast with or without a pyrenoid.   *Coccomyxa* 45

37b   Cells with a cup-shaped chloroplast and a basal pyrenoid, spherical to spheroidal.   *Palmella* 47

38a   Cells sometimes with an orange or reddish colour, reproducing by both autospores and zoospores.   *Chlorococcum* 40

38b   Cells reproducing solely by autospores.   *Chlorella* 32

PLATE 32

1 *Hydrodictyon reticulatum* R. Medway, Kent, part of the plant x 10, detail showing chloroplasts and pyrenoids x 100, 2 *Pediastrum clathratum* x 500, 3 *P. braunii,* x 700, 4 *P. biradiatum,* x 450, 5 *Chodatella ciliata,* x 950, 6 *Micractinium pusillum,* Tonbridge, Kent, x 1100, 7 *Franceia ovalis,* x 1900, 8 *Sorastrum spinulosum,* x 1300, 9 *Ankistrodesmus gracilis,* x 1300, 10 *Quadrigula pfitzeri* with autospore, Tingwall Lock, Shetland, x 1100.

(.75)

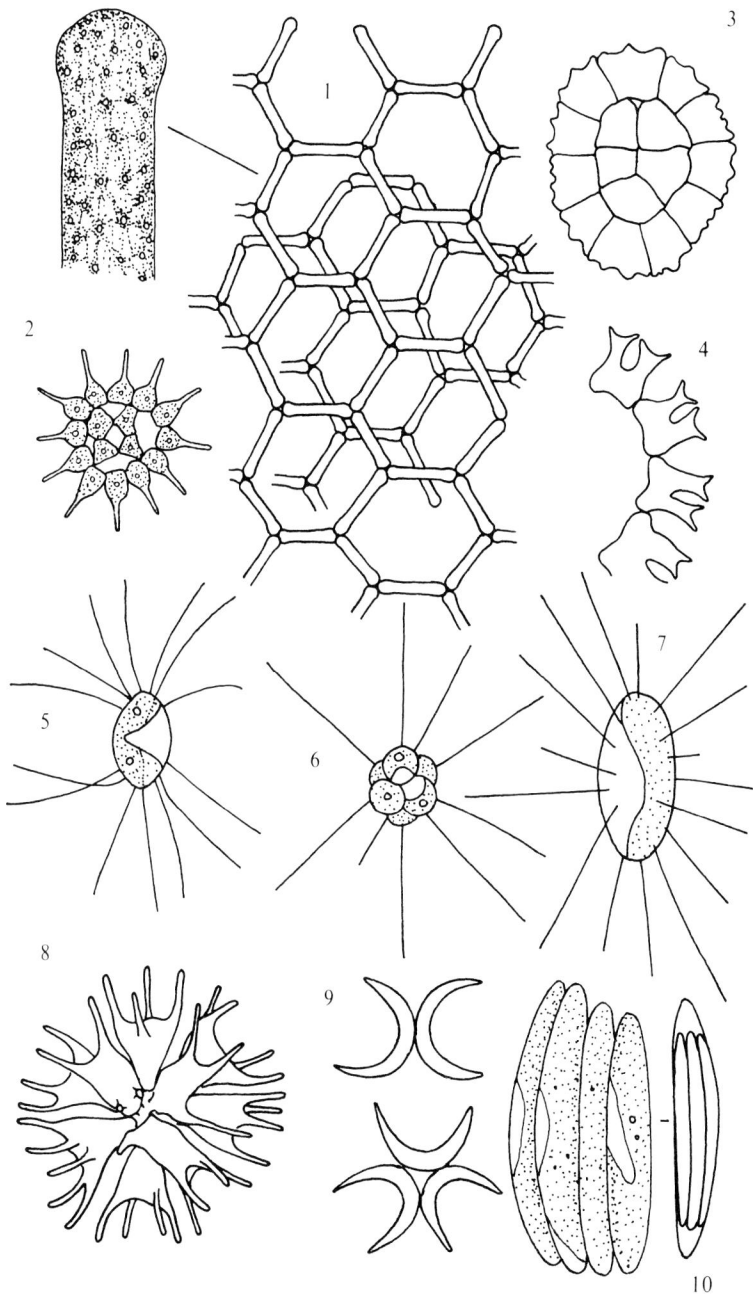

155

21. Pediastrum

The disc-shaped colonies of *Pediastrum* are a fine sight under the microscope and may be found in collections taken from lakes and small ponds where the water is rich in nutrients, particularly nitrogen. The species are usually found floating among aquatic plants but rarely found in deep water. The cell walls are tough and extremely durable and may be found in lake sediments. The individual cells measure 5-30 µm in width and there are usually (4)6-128 cells per colony. A key to the best known species is given below. In Britain, *P. boryanum* and *P. tetras* are seen most frequently.

1a   Marginal cells with a single spine 5-20 µm long.   2

1b   Marginal cells with 2-4 spines or protuberances.   3

2a   Colonies solid, of 4 or more cells, cell wall often minutely granular.   *P. simplex* (Plate 33 Fig. 1)

2b   Colonies sieve-like (rare).   *P. clathratum* (Plate 32 Fig. 2)

3a   Marginal cells with two simple protuberances, spines or notches.   4

3b   Marginal cells with 3-4 apical processes, sometimes in two pairs.   6

4a   Disc solid, of 4-many cells.   5

4b   Disc sieve-like, with about 16-128 cells, frequent.
    P. duplex (Plate 33 Fig. 4)

5a   Colonies with a cruciate appearance, of 4-8 cells, sometimes with a central hexagonal or heptagonal cell, outer cells usually notched. (cf. *Crucigenia*).
    P. tetras (Plate 34 Fig. 2)

5b   Colony of more than 8 cells, apices of marginal cells narrowed to fine points or with small terminal inflations, cell walls often granular.   *P. boryanum* (Plate 33 Fig. 3)

6a   Disc solid with 3-4 stumpy teeth or undulations on the marginal cells.   *P. braunii* (Plate 32 Fig. 3)

6b   Disc sieve-like with two pairs of spines on each marginal
      cell (rare).   *P. biradiatum* (Plate 32 Fig. 4)

22. Sorastrum

This genus is related to *Pediastrum* but the cells are arranged
around a central point where they are attached to each other by a
mucilaginous protuberance. The cell walls are minutely granular
and possess 1-4 radiating teeth or spines 4-15 μm long. The
species are rare and occur in similar habitats to *Pediastrum* (Id:
number and length of spines, cell shape; Aut, 10).

S. spinulosum is occasionally seen with colonies of kidney
shaped cells 8-25 μm long each with four spines 4-8 μm long
with 16-64 cells per colony. (Plate 32 Fig. 8).

23. Hydrodictyon

This is the 'Water Net' which is unlikely to be mistaken for
any other alga. The large multinucleate cells are attached at their
slightly inflated ends in threes to produce a fine green mesh
visible to the naked eye. The cells which are 1-3 and 4-6 ltb.
contain a network of parietal chloroplasts containing many
pyrenoids.

H. reticulatum is the only British species forming nets up to
20 cm across in eutrophic ponds and rivers, mainly in the south.
At some sites its occurrence is regular but numbers vary
considerably from year to year (Plate 32 Fig. 1).

24. Chodatella

Cells spheroidal or lemon-shaped with several long, fine spines
arranged mostly around the poles. The cells measure 8-15 μm
long, 1.5-2.5 ltb with about 3-7 spines near each pole. The single
chloroplast is parietal with a pyrenoid. The species are frequent in
the plankton of nutrient-rich lakes but easily overlooked because
of their small size. (Id: cell shape, number length and position of
spines; Aut, 0; 10).

C. ciliata (Plate 32 Fig. 5) has a tuft of 5-7 spines 12-25 μm in
length and cells 10-20 μm long, 1.5-2 ltb. *Lagerheimia* (Plate 41
Fig. 3) is a closely related genus where the spines number four

arranged in a cruciate manner, the species of which are often found with *Chodatella.*

25. Franceia

Cells spheroidal with fine spines distributed regularly over the entire cell surface. The cells are 8-20 µm long with up to 20 spines per cell measuring 5-30 µm in length. None of the species, which are all planktonic, is common. (Id: cell shape, spine structure; 4).

F. ovalis (Plate 32 Fig. 7) has cells 13-17 µm in length with 1-2 parietal chloroplasts and spines 15-23 µm long.

26. Micractinium

Cells spherical or spheroidal, in small colonies normally in groups of four. Each cell has 2-5 fine, radiating spines and a parietal chloroplast with a pyrenoid. The species are sometimes abundant in eutrophic ponds, lakes and rivers. (Id: cell shape, presence of mother cell walls, number of spines; Aut, 0; 4).

M. pusillum is the commonest form with spherical cells 3-8 µm wide and 2-5 spines 20-35 µm in length. Small spiny oospores are rarely observed among the cells in the colony (Plate 32 Fig. 6).

27. Ankistrodesmus

Cells spindle- or needle-shaped and often bent, occurring in irregular bundles and measuring (12)16-25(70) µm in length, (5)6-20 ltb. There is a single, long parietal chloroplast, usually with a pyrenoid. These algae are often abundant in eutrophic lowland waters where they occur in the plankton. Many 'species' have been described based upon the cell shape and colony form but there appear to be only three fairly well defined forms in British waters. Reproduction is by autospores. The genus *Selenastrum* is now regarded as a synonym of *Ankistrodesmus*, one species of which has been used widely as a bioassay organism to test for water quality. A related and common genus is *Monoraphidium* where the sickle-shaped cells are separate and do not form colonies. A pyrenoid may or may not be present in the single chloroplast.

1a   Cells needle-shaped, twisted or straight, ltb 20 or more, very common.   *A. falcatus* (Plate 33 Fig. 9)

1b   Cells fusiform, straight or bent, ltb 10, often much less. 2

2a   Cells S-shaped, 15-20 µm long, ltb 3-6. *A. convolutus* (Plate 33 Fig. 10)

2b   Cells sickle-shaped, 20-30 µm long, ltb 5-7. *A. gracilis* (Plate 33 Fig. 9)

28. Closteriopsis

Cells needle-shaped and 150-650 µm in length, 20-40 ltb, with a single parietal chloroplast with a row of 6-14 pyrenoids. These algae have a superficial resemblance to the desmid *Closterium* but they lack the apical vacuoles, double chloroplast and method of reproduction characteristic of that genus. The species occur in nutrient-rich ponds and lakes and are sometimes abundant. (Id: number of pyrenoids; Aut; 3).

C. longissima is a large species with 8-12 pyrenoids and 150-400(570) µm in length, 30-40 ltb (Plate 35 Fig. 4).

29. Kirchneriella

Cells sickle-shaped in small mucilaginous colonies and with a parietal chloroplast containing a pyrenoid. The cells are moderately broad and often so strongly bent as to appear almost circular. (Id: cell shape; Aut; 10).

K. obesa is a frequent British species (Plate 35 Fig. 1) with strongly bent, sausage-shaped cells 6-16 µm across. The species resemble *Ankistrodesmus* but are always found in a broad mucilaginous investment. Another closely related genus is *Nephrocytium* which only differs in the less strongly bent, often kidney-shaped cells (Plate 35 Fig. 15).

30. Quadrigula

Cells elongate-fusiform, 4-12 ltb, packed into parallel groups of 2-8 and enclosed by a clear, mucilaginous membrane. The chloroplast is parietal, usually with a pyrenoid (Id: cell shape; Aut; 4).

Q. pfitzeri (Plate 32 Fig. 10) has cells 35-50 μm in length, 5-10 ltb with rounded apices and is occasional in the plankton of softwater lakes.

31. Oocystis

Cells spheroidal or egg-shaped, usually in colonies of 2-16 enclosed by the mother cell wall. The chloroplasts are usually parietal and disc-shaped numbering 1-5(60) per cell, with or without pyrenoids. The cells range from (7)12-35(40) μm in length, 1.4-3 ltb. The species are widespread, particularly in softwater ponds and lakes. A key to many of the commoner species is given below. (Id: cell shape and thickenings, number of chloroplasts, colony form; Aut; 25).

1a   Cells lemon-shaped with two polar thickenings, chloroplasts 4-25 per cell.   2

1b   Cells spheroidal or egg-shaped, without polar thickenings.   3

2a   Chloroplasts 4-10, mother cells with 2-4(8) daughter cells. Cells 14-26 μm long.   *O. crassa* (Plate 35 Fig. 5)

2b   Chloroplasts 12-25, cells mostly solitary and 7-20 μm long, about 1.6 ltb.   *O. solitaria*

3a   Chloroplasts 1-3. Cells narrowly ovoid with pointed poles, 6-16 μm long, about 2 ltb.   *O. parva* (Plate 35 Fig. 6)

3b   Chloroplasts 4-20 per cell.   4

PLATE 33

1 *Pediastrum simplex*, x 400, 2 *P. tetras*, Hever, Kent, x 1500, 3 *P. boryanum*, Wandsworth, London, x 750, 4 *P. duplex*, Wandsworth, London, x 950, 5 *Crucigenia quadrata*, Midhurst, Sussex, x 2250, 6 *Scenedesmus quadricauda*, Southborough, Kent, x 1200, 7 *S. acuminatus*, Wandsworth, London, x 500, 8 *Coelastrum microporum*, E. Grinstead, Sussex, x 1600, 9 *Ankistrodesmus falcatus*, Southborough, Kent, x 1300, 10 *A. convolutus*, Midhurst, Sussex, x 1500, 11 *Asterococcus superbus*, Malham Fen, Yorks, x 350, 12 *Sphaerocystis shroeteri*, Medway, Kent, part of colony, x 1700.

(.75)

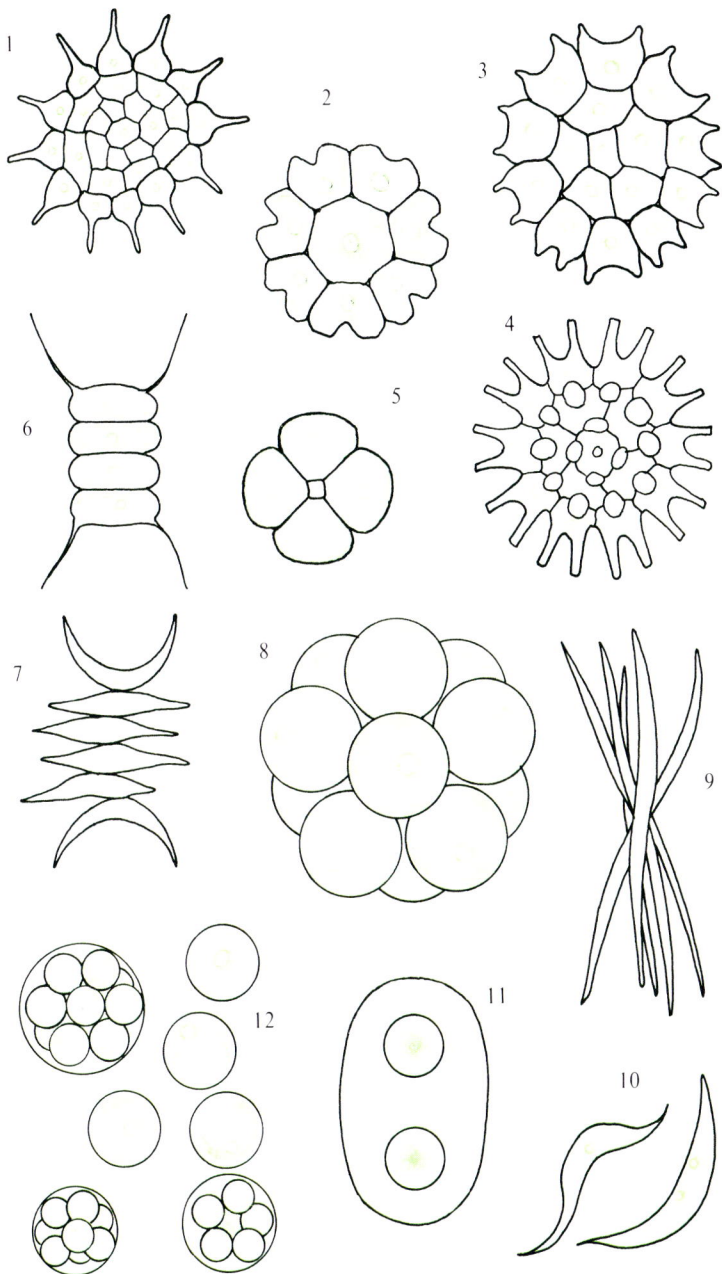

161

4a   Chloroplasts 4-8 stellate plates, cells 23-38 μm long, about 1.5 ltb.   *O. natans* (Plate 35 Fig. 7)

4b   Chloroplasts 10-20 parietal discs.   5

5a   Cells 40-50 μm long, mother cell wall smooth and symmetrical.   *O. gigas* (Plate 35 Fig. 8)

5b   Cells 15-25 μm long, about 2 ltb, mother cell wall irregular.   *O. elliptica* (Plate 35 Fig. 9)

32. Chlorella

Cells spherical or spheroidal with a single parietal chloroplast and usually a pyrenoid. The cells measure (1)2.5-12(20) μm in length. *Chlorella* is sometimes frequent in small eutrophic water bodies and some strains can be grown easily in the laboratory where they have been used extensively in research. Algal physiologists have distinguished many strains of *Chlorella* but their approach towards the classification of the species is hardly acceptable to most ecologists. The cells should be observed carefully in a culture medium to ensure that reproduction only by autospores is occurring. (Id: cell shape, cell wall thickness, chloroplast structure, presence of pyrenoids; Aut; 10).

C. vulgaris (Plate 35 Fig. 13) is the best known species with thin walled cells 2.5-10 μm in diameter, containing an open, cup-shaped chloroplast with a pyrenoid. Some species live as symbionts within aquatic invertebrates.

33. Eremosphaera

There is a single British species, *E. viridis* (Plate 35 Fig. 14) which is found occasionally among wet *Sphagnum*, usually associated with desmids. The cells are spherical and contain numerous, somewhat angular or rounded parietal chloroplasts each with a pyrenoid. The cells measure 50-200 μm in diameter.

34. Trochiscia

Cells soltary with a dense, axile chloroplast and usually a central pyrenoid. The cell walls are generally thickened, about 2-4 μm wide and either angular or with minutely granular ornamentation. The cells measure 15-40 µm in diameter and are widespread but rarely abundant on damp soil or rock, or aquatic. (Id: cell wall structure; 25).

T. reticularis (Plate 35 Fig. 10) has cells 24-32 µm wide and a thickened, regularly polygonal cell wall and is found occasionally on damp soil or rock. Some species may be spores or resting-stages of other algae.

35. Scottiella

Cell spheroidal, with prominent helical or irregular longitudinal ridges and a single parietal chloroplast with a pyrenoid. Some species are coloured red by a carotenoid pigment and occur as snow algae whilst others are known from damp rocks in northern districts. They are rarely planktonic and the species are poorly defined. (Id: details of ornamentation; 10).

S. spinosa has verrucose ridges and cells measuring 20-48 µm in length, about 2 ltb. The species are rare in Britain (Plate 36 Fig. 11).

36. Chlorochytrium

These are large, unicellular algae which grow within the leaves of aquatic flowering plants. The cells are usually smooth in outline but often somewhat irregular in shape and they contain a large and extremely dense chloroplast containing normally 1-several pyrenoids. The cells do not appear to be parasitic and probably develop within the spaces between the host cells. (Id: plant colonized; z, I; 10).

C. lemnae is perhaps the commonest species which grows within duckweed (*Lemna* spp.) with cells 50-100 µm. (Plate 35 Fig. 3).

37. Characium

Species of *Characium* are common epiphytes and occur mainly on other filamentous algae. The cells are attached by a pad of mucilage which is normally stalked. Each cell contains a single parietal chloroplast and pyrenoid and the cells range from spherical to fusiform with an acute apex. The cells are 15-80 µm in length and some species may simply be the germinating spore of other algae. (Id: cell shape; z; 30).

C. sieboldii is a frequent species with broadly spheroidal cells and a poorly defined stalk. The cells measure 40-70 μm in length, 2.5-3.5 ltb (Plate 36 Fig. 2). *Pseudococcomyxa* is a similar form but reproduces by autospores.

38. Schroederia

Cells needle-shaped and frequently arcuate, drawn into long, narrow spines at each pole and containing a single chloroplast with 1-several pyrenoids. There are few species and they often occur with *Ankistrodesmus* which they strongly resemble although they are not so common. (Id: cell shape; z; 4).

S. setigera (Plate 34 Fig. 7) is a frequent species with cells 75-90 μm long, 20-25 ltb including spines. A similar genus is *Ankyra* where one of the spines is forked. The cells of these species are composed of two halves which separate during reproduction.

39. Tetraedron

Cells angular, with or without sharp spines at the angles, 12-50 μm wide. Chloroplast axile with 1-several pyrenoids. The species are sometimes found in small numbers in the plankton or on damp rock and soil and in some forms the cell walls are thickened, particularly at the angles. A considerable number of species have been transferred to the genus *Tetraedriella* placed in the Xanthophyceae on account of the negative starch reaction. (Id: cell shape; Aut; 5).

PLATE 34

1 *Scenedesmus acutus,* x 1200, 2 *S. arcuatus,* x 750, 3 *S. armatus,* x 1000, 4 *S. spinosus,* Matfield, Kent, x 1300, 5 *S. denticulatus,* x 1600, 6 *Actinastrum hantzschii,* x 1500, 7 *Schroederia setigera,* x 1100, 8 *Crucigenia fenestrata,* Midhurst, Sussex, x 2000, 9 *Coelastrum reticulatum,* x 1300, 10 *C. cambricum,* x 1000, 11 *Coccomyxa dispar,* R. Duddon, Cumbria, x 700, 12 *Botryococcus braunii,* Wise Ean Tarn, Cumbria, x 330, 13 *Elakatothrix gelatinosa,* colony x 300, two cells x 1400, 14 *Dictyosphaerium ehrenbergianum,* Speldhurst, Kent, x 1500, 15 *Pseudosphaerocystis planktonica,* Windermere, Cumbria. x 600.

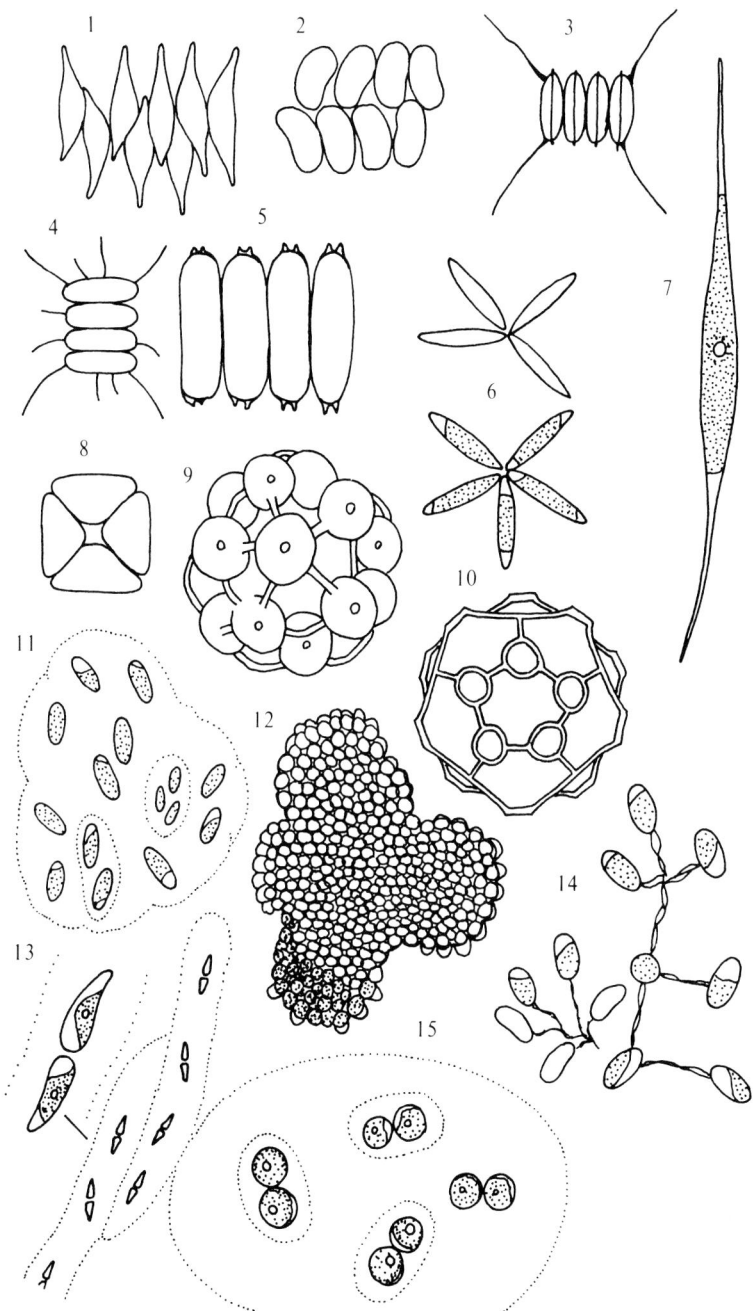

T. minimum (Plate 35 Fig. 12) is a frequent lowland planktonic form with four-angled, tetrahedral cells and slight corner thickenings. The cells measure 7-16 μm across. *Euastropsis richteri* is an uncommon planktonic alga which is rather similar to *Tetraedron* but is composed of two cells joined together in the form of two W's and measuring 10-40 μm across (Plate 41 Fig. 6).

40. Chlorococcum

Cells spherical, containing a parietal chloroplast almost completely surrounding the cell, with 1-many pyrenoids. Some species have an orange-red colour which may also be seen in *Chlorella*. The cells measure 8-40(75) μm wide and occur frequently on damp rocks, soils and eutrophic ponds but they are difficult to identify. (Id: type of zoospore development which is by successive or simultaneous cleavage, pigmentation, cell wall thickeness, abundance of vacuoles, number of nuclei; Aut; z; 10).

C. infusionum (Plate 35 Fig. 16) appears to be a common form with uninucleate cells 10-15 μm wide. Zoospore cleavage is simultaneous, and the cells are green in colour.

41. Scenedesmus

Cells in small colonies with or without regularly arranged spines on the two end cells. Each colony consists of a single, or occasionally double row of (2)4-16 cells, the walls of which are often minutely granular or ridged. The spines appear to function as flotation aids and may develop in response to changing buoyancy conditions. The species are abundant in all kinds of water but mainly in eutrophic ponds, often among vegetation. Many 'species' have been described although several studies have shown that there is great variation within individual clones so that any classification based on morphology is bound to be tentative. A key to the commoner forms is given below. *S. quadricauda* is the most frequently observed species.

1a   Cell walls appearing smooth, without warts or spines.   2

1b   Cell walls with long or short spines, walls often minutely granular.   4

| 2a | Cells fusiform with acute apices, cells often bent. | 3 |

2b Cells bean-shaped and slightly bent, c.6-12 µm long, often in double rows.   *S. arcuatus* (Plate 34 Fig. 2).

3a All of the cells the same shape, 8-20 µm long, 3-9 µm wide.   *S. acutus* (Plate 34 Fig. 1)

3b Cells at the end strongly arcuate and differing in shape from the other cells, 12-25 µm long.   *S. acuminatus* (Plate 33 Fig. 7)

4a Each cell with a prominent longitudinal ridge often ending in a short apical spine and with a further pair of spines on the outer cell. Cells 8-12 µm long.   *S. armatus* (Plate 34 Fig. 3)

4b Cells without ridges.   5

5a Cells with two (occ. 1) short (0.5-1µm) spines at each cell apex, cells elongate-spheroidal, 10-15 µm long.   *S. denticulatus* (Plate 34 Fig. 5).

5b Outer cells with at least two spines 5-15 µm long.   6

6a Cells with two prominent spines at both ends, cells 8-30 µm long, usually in groups of 4.   *S. quadricauda* (Plate 33 Fig. 6)

6b Cells with two prominent end spines plus 1-2 shorter median spines on the end- and occasionally other cells. Cells 8-13 µm long.   *S. spinosus* (Plate 34 Fig. 4).

42. Actinastrum

Cells in small star-shaped colonies, fusiform with slightly truncated or rounded ends. Chloroplast parietal with a pyrenoid. (Id: cell shape; Aut; 7).

A. hantzschii (Plate 35 Fig. 6) has cells 10-25 µm long, 4-8 ltb and is sometimes common in rivers and aquaria.

43. Crucigenia

Colonies consisting of four cells arranged in a cruciate manner so that a gap sometimes appears at the colony centre. Each cell contains a parietal chloroplast and minute pyrenoid. The cells are angular or spheroidal in shape and measure 4-10 µm in length. The species reproduce by autospores and measure 4-10 µm in length. The species reproduce by autospores and some authorities believe that the orientation of the young autospores within the parent colonies is of taxonomic significance. (Id: cell shape, thickenings in the cell wall, presence and shape of central pore; Aut; 10).

The species occur in the lake plankton but they are easily overlooked. *C. quadrata* (Plate 33 Fig. 5) is a common form with spherical or spheroidal cells 5-10 µm in length, ltb 1-1.4 with a central opening. *C. fenestrata* (Plate 34 Fig. 8) also has a central opening but the cells are broadly triangular so that the colony appears to have straight outer edges. A related genus is *Tetrastrum* where each cell has three or more radiating spines or teeth (Plate 41 Fig. 7).

44. Coelastrum

Colonies composed of closely and regularly packed spherical or spheroidal cells, each cell with a parietal chloroplast and pyrenoid. The cells measure (3)8-20(35) µm wide and the colonies are about 30-60(140) µm in diameter consisting of 8-64 cells. The species are found mainly at the edges of lakes in hilly districts but may also occur in the plankton of eutrophic lakes and rivers. (Id: cell shape and ornamentation, presence of connecting structures between the cells; Aut; 20).

C. microporum is a species with 8-32 smooth, spherical cells per colony, cells 3-17 µm wide, colonies 12-42 µm wide, without

---

PLATE 35

1 *Kirchneriella obesa*, Pevensey, Sussex, x 1100, 2 *Characium sieboldii*, x 1500, 3 *Chlorochytrium lemnae*, Cuckmere, Sussex, x 500, 4 *Closteriopsis longissima*, Wandsworth, London, x 800, 5 *Oocystis crassa*, x 1100, 6 *O. parva*, x 1500, 7 *O. natans*, x 800, 8 *O. gigas*, x 500, 9 *O. elliptica*, x 450, 10 *Trochiscia reticularis*, x 1100, 11 *Scottiella spinosa*, x 750, 12 *Tetraedron minimum*, x 2200, 13 *Chlorella vulgaris*, Tunbridge Wells, Kent, x 2200, 14 *Eremosphaera viridis*, Robinson's Tarn, Cumbria, x 300, 15 *Nephrocytium lunatum*, Malham, Yorks, x 750, 16 *Chlorococcum infusionum*, x 1900, 17 *C. wimmeri*, zoospore, x 2500.

(.75)

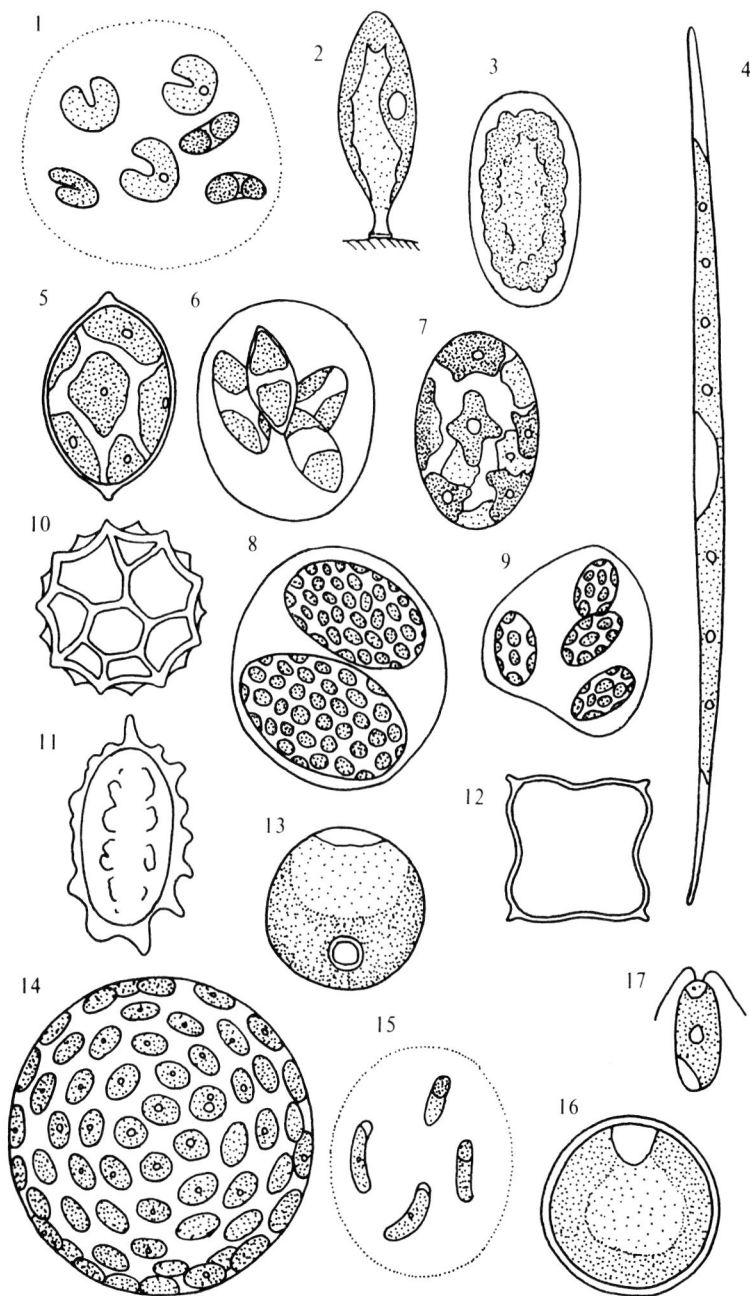

169

connecting structures. It is frequent in the plankton (Plate 33 Fig. 8). Two species with gelatinous connecting processes are *C. reticulatum* and *C. cambricum* (Plate 34 Figs. 9 and 10). In the former the processes are long and narrow and the cells are smooth and spherical, 3-10 μm wide. The latter species has much shorter processes and the cells, which measure 9-30 μm, have prominent apical thickenings.

45. Coccomyxa

Cells usually spheroidal or shortly fusiform with a parietal chloroplast with or without pyrenoids, in large gelatinous masses. The mucilage is sometimes lamellate and the cells are evenly and widely dispersed within it.

There appears to be only one well defined free-living species, *C. dispar* (Plate 34 Fig. 11). The cells measure 8-15 μm in length, 1.5-3.5 ltb and often appear in loose pairs. The chloroplast is often without a pyrenoid and reproduction is normally achieved by simple fission although reproduction by autospores has also been observed. It is often abundant on damp, base-poor rocks. Other species are the symbionts of lichens.

46. Elakatothrix

Cells fusiform or wedge-shaped, normally in pairs and attached loosely end to end and in large, somewhat filamentous, mucilaginous colonies. Chloroplasts parietal, with or without a pyrenoid. Cells sometimes with contractile vacuoles. (Id: cell and colony shape, chloroplast; 8).

E. gelatinosa is sometimes found in the plankton with narrow spindle-shaped cells 16-25 μm long, 2.8-3.3 ltb (Plate 34 Fig. 13).

47. Palmella

Cells spheroidal, spherical or rounded-cylindrical with a cup-shaped chloroplast and basal pyrenoid, in large gelatinous colonies among damp rocks and vegetation. The cells measure 5-20 μm in length and represent non-motile stages of *Chlamydomonas* of similar algae, to which reference should be made for illustration.

48. Sphaerocystis

Cells spherical, 7-20 µm in wide and numbering 4-50 in mucilaginous, approximately spherical colonies. Each cell has a parietal chloroplast containing 0-several pyrenoids and the cells subdivide to form 8-16 densely packed, botryoidal autospores giving the colonies a characteristic appearance.

Two species occur in the plankton and the first is often abundant in mesotrophic lakes. This is *S. schroeteri* with the chloroplast containing a single, large pyrenoid (Plate 33 Fig. 12). *S. sphaerocystiformis* is similar but pyrenoids are absent.

49. Pseudosphaerocystis

This genus differs from the last in the occasional presence of two small contractile vacuoles and the absence of botryoidal masses of autospores. The cells are spherical, 8-15 µm wide and occur in small groups in large mucilaginous free-floating colonies. The chloroplast is cup-shaped with a basal pyrenoid but pseudocilia are absent (cf. *Paulschulzia*). (Syn. *Gemellicystis*).

P. planktonica is a frequent British species where the cells usually occur in groups of 2 or 4, 6-10 µm wide. The colonies are up to 200 µm wide and the mucilage is frequently stratified (Plate 34 Fig. 15).

50. Asterococcus

Cells spherical, in small groups enclosed by a broad and usually lamellate mucilage envelope. Chloroplast dense and axile with a central pyrenoid and two contractile vacuoles are normally present.

A. superbus is probably the only species with cells 10-30 µm wide and a mucilage envelope up to 15 µm thick. It is often common in bog pools among *Sphagnum* (Plate 33 Fig. 11).

51. Botryococcus

The colonies of *Botryococcus* consist of cells so densely packed that they often appear almost black in colour. The individual cells are difficult to distinguish and are angular or spherical, 7-10 µm wide with a parietal chloroplast and pyrenoid. The cells are held together by a dense oily mucilage which is often brown in colour. When in the plankton net the colonies can look like small particles of peat. *B. braunii* is the only species and is widespread in the plankton of all kinds of lakes, with perhaps a preference for oligotrophic waters. The colonies grow to 1 mm in diameter. In one Australian lake the cells aggregate on the shore to form large rubbery deposits (Plate 34 Fig. 12).

52. Dictyosphaerium

Colonies composed of groups of four cells connected to each other by the remains of old mother cell walls. Cells spherical or kidney-shaped, 6-12 µm long with a parietal chloroplast and pyrenoid. (Id: cell shape; Aut, z, 0; 10).

D. ehrenbergianum is a frequent planktonic species of lakes and ponds and has spherical cells 6-10 µm wide (Plate 34 Fig. 14). The genus *Westella* should probably be regarded as a synonym of *Dictyosphaerium.*

53. Radiococcus

Cells spherical in tetrahedral, or occasionally planar groups of four, each cell measuring 7-15 µm in diameter and enclosed within a wide mucilaginous envelope. The chloroplast is cup-shaped and usually with a pyrenoid.

There are two species and *R. nimbatus* (Plate 39 Fig. 4) is the most frequent, occurring in the plankton of lakes. The colonies contain several tetrahedrally arranged groups of cells measuring 8-15 µm wide. Reproduction is by autospores.

54. Nautococcus

These neuston-algae form yellow-green scums on small ponds during warm weather but they are rarely seen. The spherical to pyriform cells sometimes possess a bell-shaped cap which is thought to assist their buoyancy but this is unlikely since few cells form them. The presence of a non-wettable cell wall seems a more likely explanation.

N. pyriformis (Plate 41 Fig. 1) has pear-shaped or spherical cells 6-12(25) µm long with an axile chloroplast and central pyrenoid sometimes obscured by starch grains. The few species are poorly defined, and some have contractile vacuoles.

Section D. Ulotrichales

1a Species with simple, unbranched filaments 2

1b Species with foliar thalli, 1-2 cells thick. 7

2a Each long cell containing several narrow annular chloroplasts with several pyrenoids. Oogamous algae with rows of spherical, rough-walled zygospores occasionally present in some of the cells (rare). *Sphaeroplea* 55

2b Chloroplasts either a single band (with or without pyrenoids,) or closely packed or reticulate, without pyrenoids, isogamous. 3

3a Chloroplasts several per cell, without pyrenoids. Cell walls often thickened, end cells H-shaped. *Microspora* 60

3b Chloroplasts one per cell, band-like, with or without pyrenoids. 4

4a Filaments with a broad mucilaginous envelope, cells rounded and often separated from each other. *Geminella* 59

4b Filaments normally without a mucilaginous envelope, cells joined together. 5

5a Filaments readily breaking up into short lengths. Pyrenoids normally absent, cells with rounded ends, mainly terrestrial. *Stichococcus* 59

5b Filaments not breaking up, cells with truncated ends, pyrenoids present, usually aquatic. 6

6a Zoospores with four flagella. Chloroplast extending from one half to almost entirely around the cell. *Ulothrix* 56

6b Zoospores with two flagella. Chloroplast normally extending less than half way around the cell. *Hormidium* 57

7a	Small terrestrial algae of nitrophilous habitats composed of short irregular ribbons up to 1 mm wide.	*Prasiola* 61
7b	Large aquatic algae usually 10 cm or more long.	8
8a	Thalli tubular and sausage-like.	*Enteromorpha* 63
8b	Thalli delicate in irregular rosettes composed of a single layer of cells.	*Monstroma* 62

55. Sphaeroplea

Filaments 11-80 μm wide composed of long cells 4-60 ltb. Each cell contains several annular parietal chloroplasts with large pyrenoids. The species are readily recognized when fertile as the cells contain several spherical oospores with ornamented walls. (Id: oospore wall structure; 0; 6).

S. annulina (Plate 36 Fig. 6) is the best known species with cells 30-70 μm wide and a double row of spinose oospores in the unspecialized female cells. It occurs rarely in small coastal pools in Britain.

56. Ulothrix

Cells (0.2)0.7-1.3(6) ltb and (2)5-20(50) μm wide, each with a single band-shaped chloroplast containing 1-several pyrenoids. The number of species is uncertain and none of the proposed schemes of classification is satisfactory.

U. zonata (Plate 36 Fig. 11) is fairly well defined with filaments (7)11-37(70) μm wide and cells 0.2-2(6) ltb. There are 1-8 pyrenoids in each chloroplast and 2-16 zoospores are produced per cell. Sexual reproduction is isogamous and the cells producing the gametes are often curved, with each cell producing

---

PLATE 36

1 *Oedogonium undulatum* with oospore and dwarf male (d), x 400, 2 *O. inframediale*, Southborough, Kent, with oospore, cap cells and dwarf males, x 300, 3 *Oedogonium* zoospore, x 600, 4 *Bulbochaete* sp., Grasmere, Cumbria, x 120, 5 *Dichotomosiphon tuberosus*, x 40, 6 *Sphaeroplea annulina*, part of a filament with oospores and annular chloroplasts, x 250, 7 *Rhizoclonium hieroglyphicum*, Tai Hu, China, x 400, 8 *Microspora spirogyroides*, x 400, 9 *M. amoena*, x 400, 10 *Stichococcus bacillaris*, Tunbridge Wells, Kent, x 750, 11 *Ulothrix zonata*, Malham, Yorks, x 750, 12 *Geminella minor*, Windermere, Cumbria, x 1000.

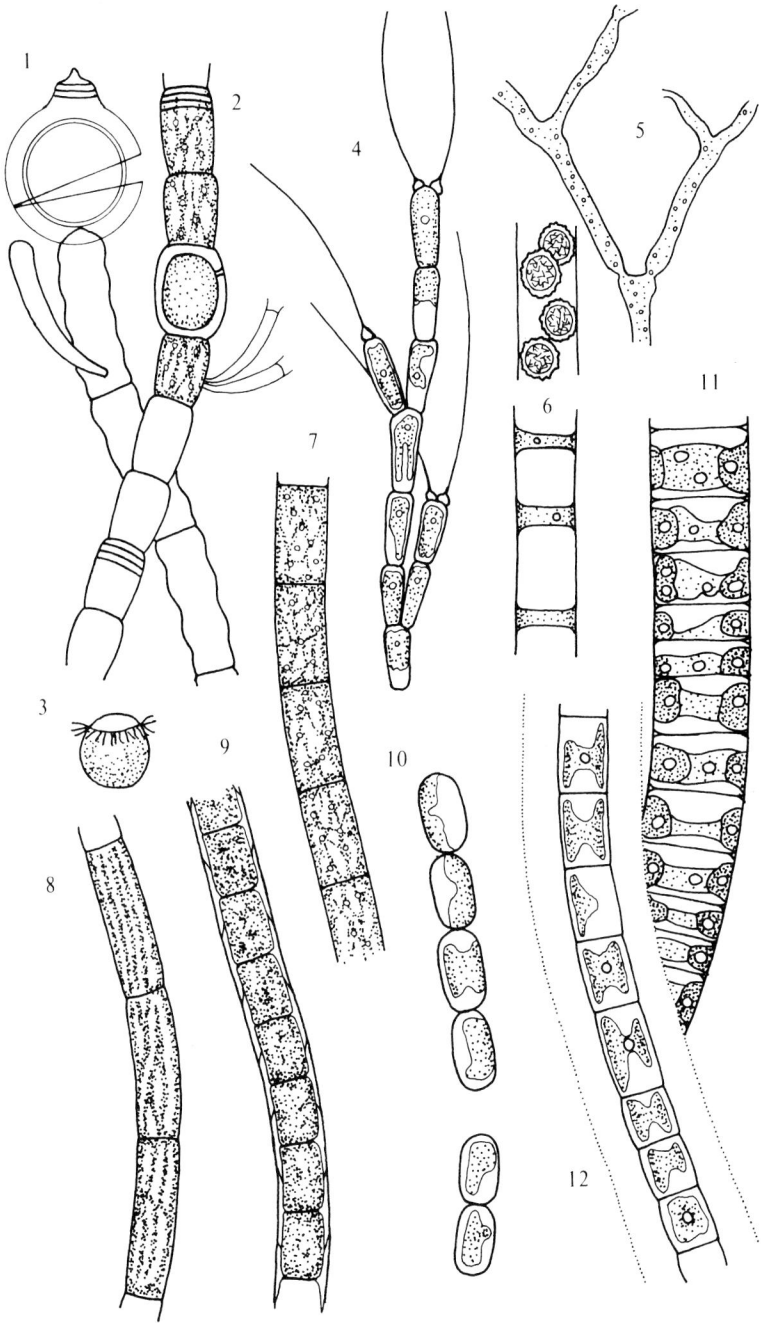

8-32 gametes. The filaments are free-floating or attached by a basal cell and small rhizoids are occasionally present.

U. mucosa is a species which has a mucilaginous investment visible when India Ink is added. The cells measure (6)10-14(16) μm wide, 0.25-2 ltb with 1-4 pyrenoids per chloroplast. Species of *Ulothrix* are often abundant in still and running shallow water.

57. Hormidium

Filaments similar to *Ulothrix*, (3)5-15(20) μm wide, cells 1-3 ltb, each with a semicircular band-shaped chloroplast and usually a single pyrenoid. The zoospores are biflagellate and can sometimes be obtained by leaving material in a dish overnight. The species are particularly common on damp soil, footpaths and in acidic, nutrient-poor streams but identification is difficult and depends upon the filament shape and width.

H. flaccidum (Plate 38 Fig. 3) is a common form with cells 6-14 μm wide and straight filaments. Some authors refer to the genus as *Klebsormidium*.

58. Geminella

The filaments of *Geminella* are frequently short with barrel-shaped cells which may be well separated from each other. Cells are (3)5-15(20) μm wide, 1-2.5 ltb each containing a band shaped, often lobed chloroplast, with or without pyrenoids. (Id: cell shape; 6).

G. minor (Plate 36 Fig. 12) has quadrate cells 4-8 μm wide, 1-1.5 ltb with usually a single pyrenoid per chloroplast. This species, in common with several others, is frequent in the plankton of lakes in hilly districts and in bog pools surrounded by *Sphagnum*.

59. Stichococcus

Filaments short and fragmenting, (3)5-10(12) μm wide, cells spheroidal, 1-3 ltb with an irregularly lobed chloroplast but no pyrenoid.

S. bacillaris (Plate 36 Fig. 10) is the only species which occurs on fertile soils, paths and the bases of walls and trees where it is often common.

60. Microspora

Filaments often in entangled masses with the broken ends H-shaped caused by a regular weakening in the middle of each cell. The chloroplasts are granular or reticulate and are frequently poorly defined. The cells measure (4)10-20(25) µm wide and 1-3(5) ltb. (Id: chloroplast structure, wall thickness and structure; z; 30). In some species the wall structure is only apparent on staining with ruthenium red or methylene blue.

M. amoena (Plate 36 Fig. 9) is a large form with cells 20-25 µm wide, 1-2 ltb and walls 2.5-8 µm thick, with a dense, ill-defined chloroplast. *M. floccosa* has a similar chloroplast but the cells measure 7-15 µm wide and the walls are thin. *M. spirogyroides* is the same size but the chloroplast is parietal and composed of longitudinal, beaded bands (Plate 36 Fig. 8). The species occur in small shallow water bodies but they are rarely abundant. A few forms appear to be tolerant of heavy metal pollution.

Cylindrocapsa is a related genus where the filaments are composed of ovoid, thick-walled cells with a dense massive chloroplast. The species are oogamous and the red or orange, swollen oogonia are a characteristic feature, with cells 10-30 µm wide and the species occur in shallow soft water sites. The cell walls have no H-end structures.

61. Prasiola

Thalli small and papery, consisting of irregular greenish strands attached to soil or stones. The cells are in a single layer and each cell is 5-20 µm wide with an axile chloroplast and pyrenoid. (Id: thallus shape; 6).

P. crispa (Plate 39 Fig. 3) is widespread in Britain with cells 8-18 µm wide forming rounded or ribbon-shaped thalli up to 1 cm in length but usually much less. The species occur in sites frequented by animals, particularly birds.

62. Monostroma

Thalli broad and delicate often in the form of several pale green, leaf-like lobes up to 15 cm across, cells spherical or angular, 8-15 µm wide, each with a parietal chloroplast and pyrenoid. There are two freshwater species which occur occasionally near the coast in shallow water.

M. membranaceum (Plate 37 Fig. 6) is the commonest species with closely packed, angular cells 7-20 µm wide. *M. bullosum* has rounded cells in regularly arranged groups, cells 6-12 µm wide.

63. Enteromorpha

Thalli tubular, dark green and either attached or free-floating in lakes, ditches and rivers. The cells develop in one or two layers (Id: thallus shape, presence of pyrenoids, layers of cells; z, I; 5).

E. intestinalis (Plate 37 Fig. 5) is the commonest species which is often abundant in southern England with large sausage-like thalli up to 1 metre long and 0.5-2 cm wide. The cells measure 10-16 µm wide and they are often irregularly arranged and not in well defined tiers. This species is also a common seaweed.

Section E. Chaetophorales

1a   Algae terrestrial, producing a green powder covering trees or rocks. Rarely filamentous and in dense, non-mucilaginous packets of spherical or angular cells, chloroplast parietal.   2

1b   Algae terrestrial or aquatic, composed of either long or short filaments, a flat disc of radiating filaments or of single cells with branched, apical hairs.   3

2a   With a small pyrenoid.   *Desmococcus* 64

PLATE 37

1 *Microthamnion kuetzingianum*, Uckfield, Sussex, habit x 300, branch x 1000, 2 *Stigeoclonium farctum* Teddington Lock, London, upright system shaded, x 250, 3 *S. tenue*, Bewl Bridge, Kent, habit x 80, cells x 300, 4 *Gongrosira incrustans*, Arncliffe, Yorks, x 600, 5 *Enteromorpha intestinalis*, Syon Park, London, plant x 1/2, cells x 400, 6 *Monostroma membranaceum*, Deal, Kent, plant x 1/2, cells x 400, 7 *Trentepohlia umbrina*, Ryde, Isle of Wight, x 500, 8 *T. calamicola*, Arncliffe, Yorks, x 225.

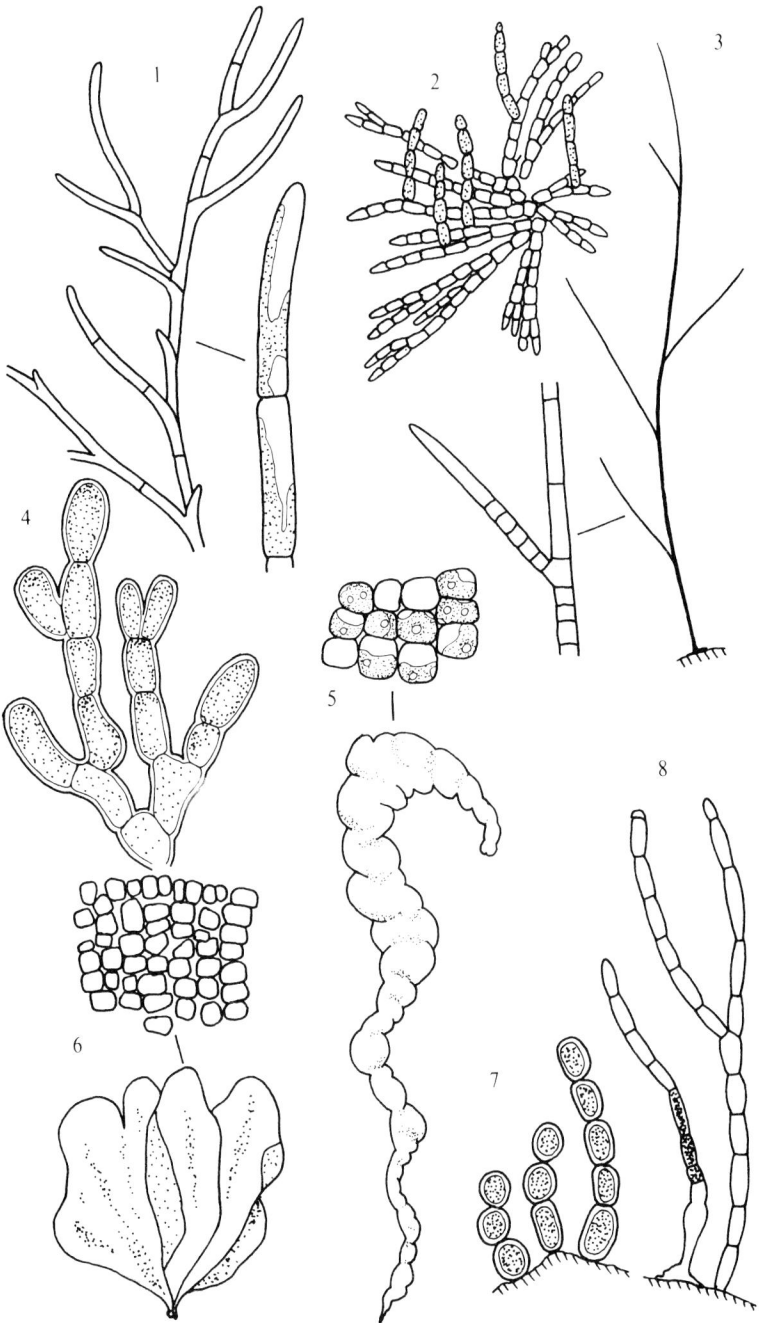

179

2b	Without a pyrenoid.	*Apatococcus* 65
3a	Cells epiphytic, solitary and with 1-2 singly forked fine hairs.	*Dicranochaete* 67
3b	Plants filamentous	4
4a	Filaments short and little branched, normally composed of less then 30 cells, some of which have 1-2 fine, erect hairs.	*Aphanochaete* 66
4b	Filaments long and branched or if short then orange in colour and without hairs, or plants composed of a flat disc of cells	5
5a	Terrestrial on rocks or the bark and leaves of trees, usually with a brown or orange colour.	6
5b	Aquatic, cells green.	7
6a	Forming small cushions or broad scurfy expanses of short or long, slightly branched filaments.	*Trentepohlia* 75
6b	Forming neat, orbicular discs up to 5 mm wide on the leaves of evergreen plants (rare).	*Phycopeltis* 76
7a	Plants in the form of a flat disc or rounded cushion closely attached to rocks or plants, cushions gelatinous and sometimes calcified.	8
7b	Plants composed of freely branched filaments attached only at their bases.	11
8a	Oogamous algae. Many of the cells with fine hairs arising from bulbous bases.	*Coleochaete* 68
8b	Not oogamous, cells without hairs.	9
9a	A flat disc, one or two cells in thickness, attached to plants or rocks.	*Protoderma* 69
9b	Plants cushion-like and gelatinous.	10
10a	Cushions up to 5 mm wide composed of slightly branched filaments with rounded or swollen apices which may serve (zoosporangia). Often heavily calcified.	*Gongrosira* 70

10b Cushions 1 mm to 4 cm wide, composed of well branched filaments with finely tapered apices. Thalli often irregular or foliar in shape and sometimes detached. *Chaetophora* 71

11a Plants consisting of a broad main axis from which regular, finely branched and tapering side branches arise.
Draparnaldia 74

11b Cells of filaments not differing greatly in size and shape. 12

12a Plants normally 1 mm or less in total length, cells 4-6 µm wide. Pyrenoids absent.   *Microthamnion* 72

12b Plants normally exceeding 1 mm in length, cells 8-25 µm wide. Pyrenoids present.   *Stigeoclonium* 73

64. Desmococcus

Cells spherical or angular and in dense, often neat packets. The chloroplast is parietal and covers about one half of the cell and contains a small pyrenoid. The cells are 5-20 µm wide and small rod-like cells are sometimes produced from rough-walled, spherical sporangia.

D. vulgaris (Plate 39, Fig. 1) is widely distributed on tree bark and fences, particularly in areas with some air pollution which reduces competition from lichens. It occasionally grows into short, irregular filaments. (Syn. *Pleurococcus*).

65. Apatococcus

The colonies are similar to *Desmococcus* but the chloroplast, which is parietal and lobate, has no pyrenoid.

A. lobatus has cells 8-14 µm wide and the cells reproduce by simple fission or zoospores. It appears to be more common in Britain than *Desmococcus* and occurs in similar habitats (Plate 39 Fig. 2).

66. Aphanochaete

These algae are small and easily overlooked and occur as epiphytes of aquatic plants, including other algae. The cells form short filaments 6-15 µm wide, each with a parietal chloroplast and pyrenoid. Some cells possess fine hairs which may be sheathed at the base. (Id: form of plant, number and structure of hairs; z; 5).

A. polychaete (Plate 39 Fig. 8) is a common species with spheroidal cells 6-10 µm wide, each with 1-4 fine erect hairs.

67. Dicranochaete

Cells 8-12 µm wide with 1-2 singly forked, apical hairs, 3-5 times as long as the cell. The chloroplast is parietal with a single pyrenoid and contractile vacuoles are sometimes present. The species are all solitary epiphytes and rarely seen.

D. reniformis (Plate 39 Fig. 5) is characterized by its flattened cells 10-35 µm d.

68. Coleochaete

Plants consisting of a flattened disc or rounded cushion of cells, some of which have long fine hairs which are sheathed at the base. The filaments are normally well branched and 8-20 µm wide. Chloroplasts parietal with a pyrenoid. The species are frequent epiphytes forming cushions up to 3 mm wide and large, spherical, corticate oogonia are often seen. (Id: colony form and structure; z, 0; 10).

There are two common British species. *C. scutata* (Plate 39 Fig. 7) is of a flat disc up to 1 mm wide composed of cells 10-23 µm wide and is rarely fertile. *C. pulvinata* (Plate 39 Fig. 6) forms gelatinous hemispherical cushions up to 5 mm across composed of branched and radiating filaments bearing hairs and frequently has numerous oogonia. The cells are 20-25 µm wide and 1-3 ltb. *C. nitellarum* is an unusual species which grows within the cells of Charophytes.

69. Protoderma

Plants consisting of small circular discs of cells 5-15 µm wide. The discs are formed of 1-several layers of cells and may become filamentous towards the margins. Chloroplasts parietal with a pyrenoid, hairs absent.

P. viride is a frequent epiphyte of aquatic plants (Plate 39 Fig.

9) but could be confused with young development in *Stigeoclonium* species with which it should be carefully compared.

70. Gongrosira

Plants forming small green cushions up to 5 mm wide on rocks, plants and old wood. The cells form branched filaments (2)4-20(25) µm wide 2-5 ltb and arise from a basal, prostrate system of cells. The apical cells are often enlarged and may serve as zoosporangia. (Id: plant habit, position of zoosporangia; z; 10).

G. incrustans is frequent in streams flowing over hard limestone and the small colonies are densely encrusted with calcium carbonate. The cells measure 5-8 µm wide, ltb 1-3 and are often enveloped in abundant mucilage (Plate 37 Fig. 4). There are several related but little known genera, some members of which bore into shells or old wood.

71. Chaetophora

Thalli large and variable in size and shape, containing radiating groups of filaments which are gracefully branched and attenuated to fine points. The filaments often die back towards the base and the entire colony is enveloped in a stiff mucilage. The cells are (4)6-15(25) µm wide and (2)4-10 ltb with a parietal, sometimes band-like chloroplast and a pyrenoid. (Id: colony shape; z, I, 6). There are two common 'species' in Britain.

C. incrassata (Plate 38 Fig. 9) has flattened, foliar thalli clearly divided towards the margin whilst *C. pisiformis* has smaller, globular thalli. The species occur attached to stones and aquatic plants at the edges of lakes and ponds, mostly in the north and west. The plants sometimes become calcified in hard waters.

72. Microthamnion

The small delicate colonies consist of narrow branched cells (2)4-6 µm wide, about 4 ltb. The species are found attached to stones or plants or occasionally free floating. (Id: branching habit; z; 3)

M. kuetzingianum (Plate 37 Fig. 1) is frequently found growing upon coarse sediments and among water plants in clean, shallow water. There are numerous, short branches and the cells are 3-5 µm wide.

73. Stigeoclonium

Filaments arising from a basal system of cells which may develop laterally to produce a broad disc of tissue. The main filaments are 8-25 µm wide with cells (0.5)2-5 ltb, each with a parietal chloroplast and pyrenoid. Branches develop either regularly or, more often, irregularly and the apices are often attenuated to fine hairs. The species are common attached to rocks, stones and water plants in shallow water, particularly in eutrophic or organically polluted sites. There appear to be a small number of variable species but numerous forms have been described. (Id: development of basal system, form of branching, development of hairs; z, I; 20).

S. tenue (Plate 37 Fig. 3) is the most widely reported form in Britain with well branched filaments and the main axis 6-15 µm wide. The branches are both opposite and alternate and the apices are acute but not drawn out into fine hairs. *S. farctum* is a form with a much reduced upright system but a well developed basal system which resembles *Protoderma* in young plants. It is usually found attached to stones and submerged plants in rivers (Plate 37 Fig. 2).

74. Draparnaldia

Plants slender, lax and gelatinous, composed of a few main axes from which tufts of side branches regularly arise. The cells of the main axis are large, measuring 20-130 µm wide and often

PLATE 38

1 *Spondylosium planum*, Ben Cruachan, Argyll, x 1500, 2 *Desmidium swartzii*, Robinson's Tarn, Cumbria, x 240, 3 *Hormidium flaccidum*, Tunbridge Wells, Kent, filament x 1600, zoospore x 2400, 4 *Pithophora oedogonia*, with akinetes, x 170, 5 *Cladophora glomerata*, R. Wandle, London, plant x 70, detail x 180, 6 *C. sauteri*, Malham Tarn, Yorks, plant x 1/2, 7 *Spirogyra* sp. (Crassa group), Midhurst, Sussex, x 200, 8 *Draparnaldia mutabilis*, Grasmere, Cumbria, x 180, 9 *Chaetophora incrassata*, R. Duddon, Cumbria, plant x 1, detail x 400.

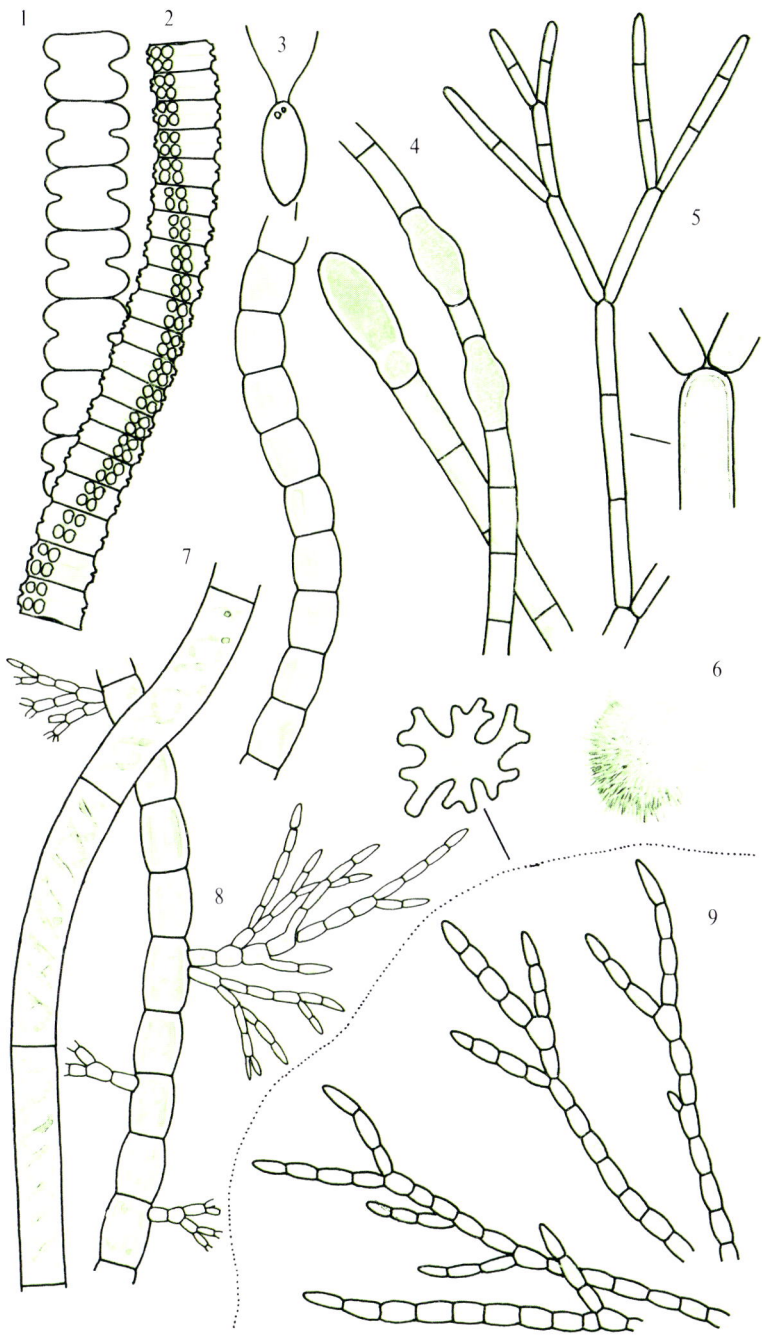

185

barrel-shaped. The side branches are much narrower and the cells are attenuated towards the apices. The chloroplasts are parietal and band-shaped or net-like, with 1-several pyrenoids. The species are epiphytes reaching up to 10 cm in length and normally found in clean, slow flowing water. (Id: type of branching; z, I; 3).

Two species are frequent in Britain. *D. mutabilis* (Plate 38 Fig. 8) has long, fine lateral branches with main axes 40-70 µm wide and the laterals develop from a prominent side branch which ends in a fine point. *D. glomerata* has shorter groups of side branches without a well developed side branch and the main axis is 50-100 µm wide. Intermediate forms are frequently found. *Draparnaldiopsis* is a similar genus not yet recorded from Britain. Here the main axis is composed of alternately long and short cells and the tufts of laterals are usually much reduced.

75. Trentepohlia

Plants consisting of small or extensive orange-brown tufts growing on sheltered rocks or trees. The filaments arise from a basal crust and they are usually branched once or twice and resemble stag's horns. The cells often have thick, faintly lamellate walls and the green chloroplasts are normally obscured by the deep orange pigmentation. Oil and not starch appears to be the main storage product and small, flask-shaped zoosporangia can frequently be found attached to the filaments. These bodies often become detached before liberating their spores. The species are abundant in the tropics but there are few forms in temperate regions and their taxonomy is much in need of revision. (Id: position of sporangia, whether sessile or on pedicels, presence of hairs, development of upright and basal system; z, I; 40). The temperate forms possess sessile sporangia and are devoid of hairs.

PLATE 39

1 *Desmococcus vulgaris*, Speldhurst, Kent, x 1100, 2 *Apatococcus lobatus*, Southborough, Kent, x 800, 3 *Prasiola crispa*, plants x 1, cells x 300, 4 *Radiococcus nimbatus*, Windermere, Cumbria. x 525, 5 *Dicranochaete reniformis*, x 1000, 6 *Coleochaete pulvinata* with corticate oospores, Grasmere, Cumbria, x 200, 7 *C. scutata*, x 550, 8 *Aphanochaete polychaete*, Ruthin, Clwyd, x 1000, 9 *Protoderma viride*, Grasmere, Cumbria, x 225.

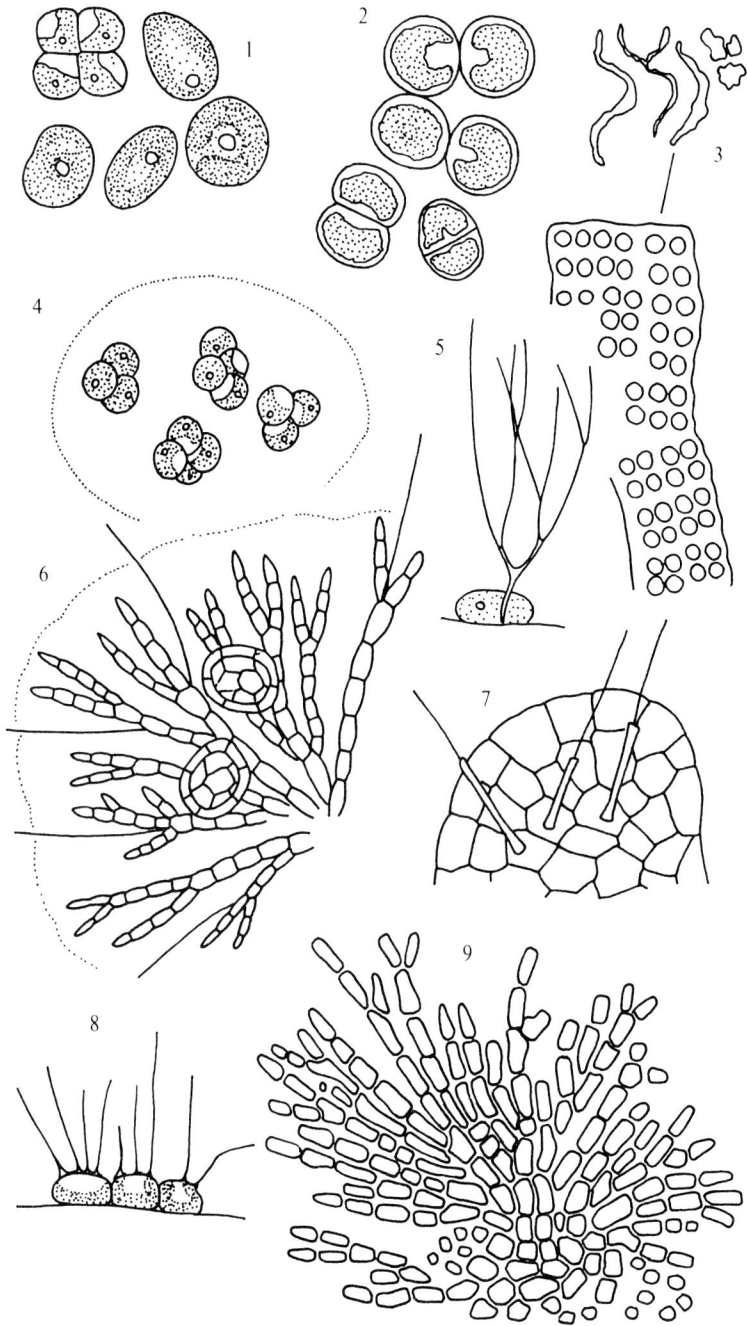

187

Key to the British species

1a   Erect filaments 5-10(12) μm wide, often slightly narrowed towards the apices. Common on rocks and trees.

T. calmicola (Plate 37 Fig. 8)

1b   Erect filaments (10)12-28(35) μm wide.   2

2a   Forming a thin pale to dark brown scurfy crust composed of short, mostly unbranched filaments with rounded cells. Abundant on shaded cement walls and gravestones. *T. umbrina* (Plate 37 Fig. 7)

2b   Forming distinct cushions or crusts 50 μm to 10 mm thick. Filaments usually long and branched composed of cylindrical cells. Sporangia often present, usually lateral and 30-45 μm wide.   *T. aurea* (Plate 14 Fig. 9)

76. Phycopeltis

Alagae forming orbicular orange patches on the leaves of evergreen trees, 1-10 mm wide. The filaments are closely packed together and radiate outwards and the plants resemble small upturned mushroom caps. The cells at the centre of the colony are less regularly arranged and angular, 8-20 μm wide. The cell contain several discoidal chloroplasts, often without pyrenoids. (Id: presence or absence of fine hairs and erect sporangia, colony structure; z, I; 10). The species are common in the tropics but rare elsewhere.

P. arundinacea (Plate 14 Fig. 8) occurs on ivy leaves in the south of Ireland forming orange thalli up to 2 mm wide. The radiating filaments are closely packed together, hairs and sporangia are absent. *Cephaleuros*, a genus not recorded from Britain forms similar discs which develop beneath the leaf cuticle and can cause losses in commercial crops.

Section F. Oedogoniales

1a   Filaments simple, without hairs.   *Oedogonium* 77

1b   Filaments branched, some cells with hairs arising from bulbous bases.   *Bulbochaete* 78

77. Oedogonium

Cells frequently barrel shaped and slightly enlarged towards one end, forming long, unbranched filaments (2)8-40(80) µm wide, cells 2-5(10) ltb. The faint annular markings are normally clearly visible towards the ends of some of the cells. The chloroplast is parietal and net-like with several prominent pyrenoids. Identification is only possible with fertile plants which are not particularly frequent. The reproduction process is involved and there appear to be several variations on the theme. In some species, *dwarf males* (Plate 37 Fig. 2) grow upon the females and then proceed to release motile gametes which have only a short distance to swim towards the eggs and achieve fertilization. In other species the male and female plants are the same size. The oogonia are easily recognized as large ovoid cells which develop thick walls after fertilization. The male cells which are multiflagellate, enter the egg through a pore and in some species a definite lid is present. This is an important character in classification, together with the ornamentation of the resulting oospore. Over 200 species have been described. These algae are abundant in stagnant or slow-flowing shallow water throughout the country. They are most frequently found fertile in midsummer. Large, multiflagellate zoospores are also occasionally produced, one from each cell (Plate 36 Fig. 3).

O. inframediale (Plate 36 Fig. 2) is a species with spherical oospores opening by a non-median pore and fertilized by dwarf males. The cells measure 16-26 µm wide. *O. undulatum* is one of the few species which can be identified when sterile (Plate 36 Fig. 1). The cells are elongate with wavy margins measuring 14-22 µm wide. When fertile, the oospores are seen to open by a lid. This species also has dwarf males.

78. Bulbochaete

Plants consisting of a series of acutely branched filaments, either attached to rocks or plants, or free-floating. The cells are frequently barrel-shaped as in *Oedogonium* and measure (8)10-30(40) µm wide, 2-4 ltb. The oogonia may develop laterally or terminally and they are often capped by a terminal hair. The hairs develop in ones or twos on many of the cells but they are most often apical. (Id: shape and structure of the oospore, ornamentation of the oospore, relation of oospore to adjacent cell, whether lateral or terminal, z; 80). The species are frequent, particularly in hard water lakes. The plant illustrated (Plate 36 Fig. 4) is sterile.

A related genus, not recorded from Britain is *Oedocladium*, with well branched filaments without hairs. The plants are sometimes found on damp earth to which they are attached by fine rhizoids.

Section G. Siphonocladales

1a   Filaments, regularly or irregularly, branched in dense tufts or cushions.   2

1b   Filaments unbranched, or at most with occasional rhizoid-like processes.   *Rhizoclonium* 80

2a   Some filaments with long, inflated, barrel-shaped and deeply pigmented akinetes (spores). Rare in Britain.
                                                                Pithophora 81

2b   Akinetes absent, common.   *Cladophora* 79

79. Cladophora

These species form coarse green mats in hardwater streams and rivers and are among the most abundant green algae. The filaments are usually dichotomously branched, at least near the apices and the cells are long and cylindrical, (10)20-80 µm wide, 5-15 ltb. Each cell contains a parietal, net-like chloroplast with numerous pyrenoids. (Id: habit, form and structure of the basal attachment cells; z, I; 20).

In Britain, there appear to be only two distinct forms. *C. glomerata* (Plate 38 Fig. 5) is by far the commonest which occurs as deep green mats or long, isolated strands up to 1 metre in length. The filaments are well branched but the degree of branching is extremely variable. It is particularly common in hard, nutrient-rich water. *C. sauteri* is a less common form which forms spherical cushions of radiating, little branched filaments on the floor of shallow calcareous lakes (Plate 38 Fig. 6). The cushions are 1-10 cm across.

80. Pithophora

These algae are similar in habit to *Cladophora* but produce large, barrel-shaped akinetes. The branching also appears to be slightly different in that the side branches often arise a little below the transverse wall of the main axis. The cells measure 20-100 µm wide with akinetes 50-400 µm long, 1-4 ltb.

P. roettleri has both barrel-shaped and cylindrical akinetes whilst *P. oedogonia* has only regular, barrel-shaped akinetes (Plate 38 Fig. 4). The last species has occasionally been noted in hot houses and thermal outfalls in Britain.

87. Rhizoclonium

Filaments simple or with occasional short and poorly developed side branches or rhizoidal outgrowths. The algae form coarse and often dense mats in shallow, hard water and the cells contain a net-like chloroplast with many pyrenoids. (Id: cell shape, filaments bent or straight; z; 6).

R. hieroglyphicum is the only British species with cells (10)20-50 µm wide, 2-8 ltb (Plate 36 Fig. 7) it is abundant and occurs in similar places to *Cladophora* to which it is closely related.

Section H. Dichotomosiphonales

88. Dichotomosiphon

Plants consisting of well branched, siphonaceous cushions or mats, usually partly immersed in the fine sediments of eutrophic, hard water lakes. The tubular filaments measure 50-100 µm in width and often bear the distinctive reproductive organs at their tips.

The only species, *D. tuberosus* (Plate 36 Fig. 5) is oogamous and the mature, yellow oospores measuring up to 300 µm across can be observed without a microscope. The chloroplasts consist of

191

numerous parietal discs, each with a pyrenoid. This interesting alga has not yet been recorded from Britain.

Section I Zygnematales

suborder Euconjugatae

1a	Chloroplasts helical, 1-14 per cell.	*Spirogyra* 83
1b	Chloroplasts band-shaped, irregular or stellate, not helical.	2
2a	Several ribbon-like chloroplasts per cell. Conjungation tubes short or absent (rare).	*Sirogonium* 84
2b	Chloroplasts one per cell if ribbon-like.	3
3a	Chloroplasts two per cell, axile and stellate with a central pyrenoid.	*Zygnema* 86
3b	Chloroplasts one per cell.	4
4a	Chloroplast a simple plane or twisted ribbon. Cell walls thin, aquatic.	*Mougeotia* 85
4b	Chloroplast axile and irregular, not ribbon-like. Cell walls normally thick and plants with a purple tinge, growing on sandy soil.	*Zygogonium* 87

83. Spirogyra

Plants often forming large deep green cushions or wefts in shallow ponds and ditches. Cells with 1-15 helical chloroplasts each containing several pyrenoids. Conjugation is most often observed in late summer when the filaments lose their deep green colour and become yellowish. Several filaments normally become aligned for part of their length and a series of bridges develop between adjacent cells. The nucleus of one cell then passes across and fuses with the other to produce a zygospore which often secretes a thick and ornamented wall. The zygospores should be examined in a 5% solution of potassium hydroxide so that the wall structure can be clearly seen. In some species the bridge is formed by only one cell whilst in others both cells cooperate in the process. The current schemes of classification are not satisfactory because the range of variation in most of the species has not been investigated. There are two common types of cross wall structure which are helpful in identification. There is the 'simple' type (Plate 38 Fig. 7) where the wall is single or biconvex and there is the 'replicate' type where a collar of wall material appears on either side of the septum (Plate 40 Fig. 1). *S. colligata* (Plate 40 Fig. 5) has a peculiar wall structure where the cells fragment and form short H-pieces. About 75% of the species have a simple septum. (Id: number of chloroplasts per cell, wall structure, presence of rhizoids, number of cells forming conjugation tube, shape colour and detailed structure of the zygospores, lateral or scalariform conjugation; 300).

The species are not often found fertile in Britain. To assist identification I have subdivided the species into four groups based upon the cross wall structure and cholorplast number.

Key to morphological groups

A   Cross walls simple.

One chloroplast per cell   *Condensata* group.

2-16 chloroplasts   *Crassa* group (Plate 40 Fig. 3)

B   Cross walls replicate

One chloroplast per cell   *Inflata* group (Plate 40 Fig. 1)

2-4 chloroplasts   *Insignis* group (Plate 40 Fig. 4)

Most of the species belong to the Condensata group. A few forms are attached to the substream by rhizoids, e.g. *S. rhizoides* (Crassa) (Plate 40 Fig. 2) and *S. affinis* (Condensata). The filaments of *Spirogyra* are (3)8-40(200) μm in width, 1-10 ltb. Their distribution and ecology has been hampered by identification difficulties.

84. Sirogonium

The cells possess 3-10 long and narrow chloroplasts which are not so strongly twisted as in *Spirogyra*. The filaments measure

(35)50-100 μm wide, (2)4-7 ltb. When fertile, the short conjugation tubes can be seen together with yellow-brown to black, smooth or ornamented zygospores. The species are rare in Britain. (Id: zygospore shape and ornamentation; 15).

S. sticticum (Plate 40 Fig. 10) occurs in fast flowing water with cells 40-55 μm wide, 2-6 ltb and 3-6 slightly twisted chloroplasts each containing several pyrenoids. The zygospores are smooth and spheroidal, yellowish-brown and up to 75 μm long. All species have simple cross walls.

85. Mougeotia

Filaments (8)15-25(60) μm wide, cells (3)5-12 ltb, containing a single, large, ribbon-shaped chloroplast and a row of pyrenoids. The choloroplast appears to occupy most of the cell when viewed face-on but one or two twists sometimes occur. The choloroplasts can change their position in response to the direction of light using muscle-like structures. The species occur in all kinds of habitats but identification can only be made with fertile material which is not commonly seen in Britain. When conjugating, the cells involved frequently become strongly flexed at the centre and the zygospore develops within the short conjugation tube. (Id: zygospore colour, shape and ornamentation, cell width, number of pyrenoids per chloroplast; 100).

M. gracillima (Plate 40 Fig. 7) has frequently been recorded from Britain with cells 5-7 μm wide and quadrate, minutely verrucose zygospores with concave sides, 20-30 μm wide. *M. parvula* has cells 8-13 μm wide and the zygospores are smooth and spherical, 13-14 μm in diameter. The cells which have conjugated are not so strongly flexed as in the previous species (Plate 40 Fig. 8).

PLATE 40

1 *Spirogyra inflata*, vegetative and conjugating cells, x 300, 2 *S. rhizoides*, filament with rhizoides x 100, cells x 400, 3 *S. nitida* (Crassa group), Tonbridge, Kent, x 300, 4 *Spirogyra* sp. (Insignis Group), Speldhurst, Kent, x 200, 5 *S. colligata*, x 200, 6 *Mougeotia* sp., Windermere, x 200, 7 *M. gracillima*, zygospore and conjugating cells, x 500, 8 *M. parvula*, zygospore and conjugating cells, x 400, 9 *Zygnema pectinatum*, vegetative cells and scalariform conjugation, x 200, 10 *Sirogonium sticticum*, x 200, 11 *Zygogonium ericetorum*, Ashdown Forest, Sussex, x 200.

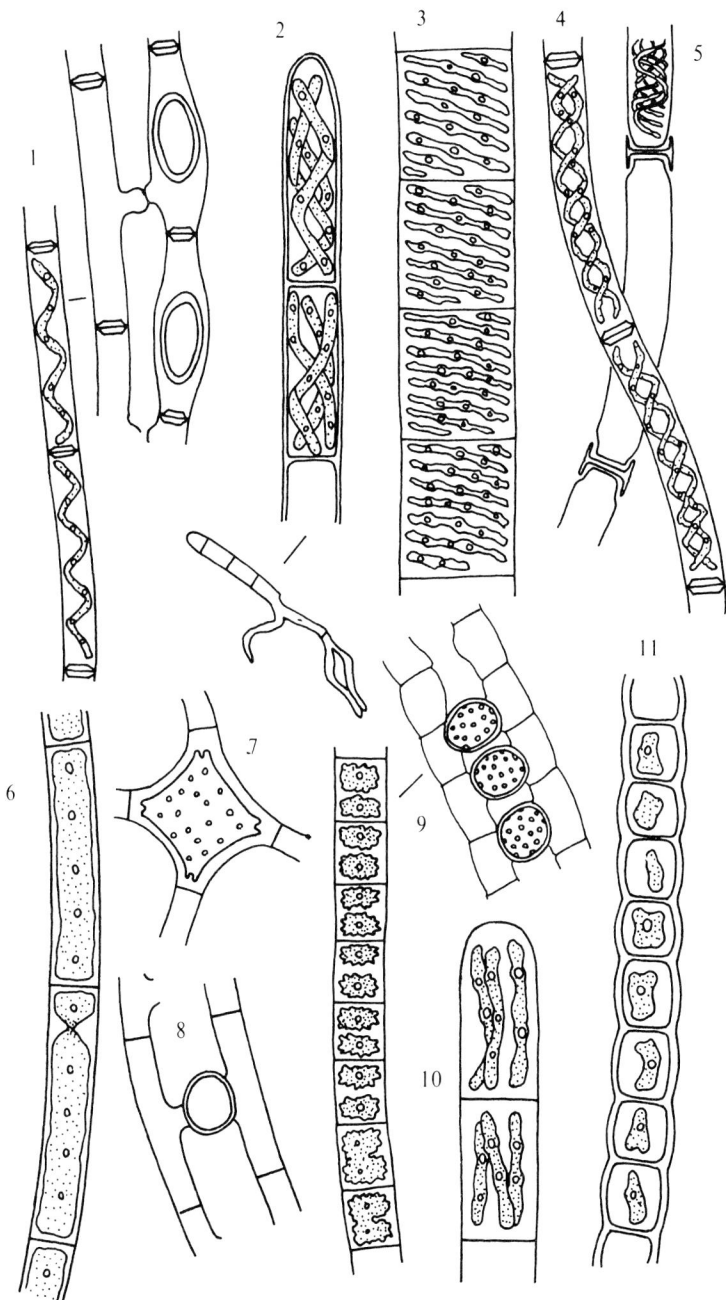

195

A related genus is *Debarya* which can only be distinguished from *Mougeotia* in fertile material, where the conjugation cells are filled with a refractive, pectic material. The species are few in number but they have been recorded from Britain. No refractive material is present in *Mougeotia*.

86. Zygnema

The species are readily distinguished by the two, normally distinctly stellate chloroplasts per cell separated by a clear area which contains the nucleus. The filaments are often of great length and measure 8-25(50) µm wide, cells 2-3(10) ltb. The cross walls are simple and the species are common throughout the country in shallow water, often occurring with other Zygnematales and *Oedogonium*, but they are rarely fertile. (Id: shape, colour and ornamentation of the zygospores, position of the zygospores within the conjugation; 100).

Z. pectinatum (Plate 40 Fig. 9) is sometimes seen fertile in Britain. The zygospores are formed within the conjugation tubes and are spheroidal, yellow-brown and covered with small pits 2-3 µm wide. The vegetative cells are 30-40 µm wide.

87. Zygogonium

In this genus, each cell contains a single, axile chloroplast which is often variable and irregular in shape and normally contains a small pyrenoid. The distinguishing character of the genus is found only in conjugating cells where a cell wall develops prior to conjugation wilting off the resulting zygospore. This character is also found in *Sirogonium* but conjugation is rarely observed in either genus.

The only British species, *Z. ericetorum* (Plate 40 Fig. 11) is readily distinguished by its greatly thickened cell walls which are often lamellate and up to 6 µm wide. The cells normally contain a purple pigment and are 8-35 µm wide, 0.8-1.5 ltb. The chloroplasts are sometimes obscured by numerous oil droplets and are extremely irregular in size and shape. This alga forms extensive purplish-brown mats on sandy soils throughout the country but in the shade the purple pigment is normally lacking.

The species is a pioneer colonizer and assists in the prevention of erosion and establishment of seedlings on recently denuded sandy soils. The zygospores are smooth and spherical and formed in the conjugation tubes.

Section I Suborder Desmidioideae

1a   Cell wall composed of a single piece, without ornamentation. Cells dividing by simple fission, median constriction absent.   (Saccoderm desmids). 2

1b   Cell wall composed of two pieces, often ornamented with pores, granules or spines. Cell division occurring through enlargement of the median constriction which is pronounced or slight.   (Placoderm desmids). 6

2a   Chloroplast a single axile ribbon or plate.   3

2b   Chloroplast axile with radiating processes.   4

3a   Cells elongate-spheroidal or shortly cylindrical with 0-2 pyrenoids per chloroplast, common.   *Mesotaenium* 88

3b   Cells elongate, slender with 3 or more pyrenoids per chloroplast, scarce.   *Roya* 89

4a   Cells elongate, slender, >6 ltb, with 1 chloroplast. *Roya* 89

4b   Cells shorter, with two chloroplasts, ltb <6.   5

5a   Chloroplasts rounded.   *Cylindrocystis* 90

5b   Chloroplasts elongate with longitudinal furrows. *Netrium* 91

6a   Cells elongate-rectangular, ltb >10, without a median constriction and with 1-2 ribbon-like chloroplasts, cells often in loose, zig-zag filaments and not clearly divided in two halves.   *Gonatozygon* 103

6b   Cells divided into two semicells.   7

7a   Unicellular or in mucilaginous colonies.   8

7b   Filamentous   18

| 8a | Unicellular | 9 |

| 8b | Cells in colonies, connected by fine mucilaginous strands, sometimes calcified, cells rounded with a median constriction. | 17 |

| 9a | Cells elongate, sometimes sickle-shaped, usually >4 ltb. Median constriction slight or absent, apical vacuoles often present. | 10 |

| 9b | Cells broad, <4 ltb, often with a complex outline and pronounced median constriction. | 13 |

| 10a | Cells sickle shaped or drawn out into spine-like apices. | *Closterium* 94 |

| 10b | Cells with rounded apices, slightly, if at all tapered. | 11 |

| 11a | Cell apices notched. | *Tetmemorus* 92 |

| 11b | Cell apices smooth. | 12 |

| 12a | Cells elongate-rectangular with truncated apices and wavy median constriction, >8 ltb. | *Pleurotaenium* 95 |

| 12b | Cells with rounded apices, <6 ltb, with slight, smooth median constriction. | *Penium* 93 |

| 13a | Cells strongly flattened, with an apical notch or incision. | 14 |

| 13b | Cells slightly flattened without an apical notch. | 15 |

| 14a | Cells with a small apical notch and a smooth, lobed outline, often with regular bulges in vertical view. | *Euastrum* 96 |

| 14b | Cells with a deep, acute notch and a well-incised star-shaped outline, without bulges in vertical view. | *Micrasterias* 97 |

| 15a | Cells with marginal spines or angular protuberances. | 16 |

| 15b | Cells without spines or protuberances, walls smooth or minutely granular. | *Cosmarium* 100 |

| 16a | Cells in vertical view with 2-8 equally spaced angles ending |

in either a rounded, thickened protuberance or drawn out into a spine. *Staurastrum* 99

16b Cells in vertical view rounded, with a rounded central protuberance on each broad face and with a regular series of spines which are isolated or in small groups. *Xanthidium* 98

17a Colonies heavily calcified, occurring in calcareous streams, rare *Oocardium* 102

17b Colonies not calcified, usually the planktonic. *Cosmocladium* 101

18a Cells cylindrical, without deep lobes, median constriction slight. 19

18b Cells usually short, with deep lobes either in face-view or in vertical view (sometimes both), median constriction usually deep. 20

19a Median constriction a small notch, cells barrel-shaped with a few transverse rows of pores. *Bambusina* 106

19b Median constriction a slight, smooth depression, cells quadrate. *Hyalotheca* 107

20a Cells elliptic in vertical view, without a central gap between adjacent cells. *Spondylosium* 105

20b Cells with (2)3-4 angles in vertical view and a small lens-shaped gap between adjacent cells. *Desmidium* 104

88. Mesotaenium

Members of this genus are commonly found in base-poor terrestrial situations. The cells are capsule-shaped and measure 20-35(100) µm long, 1.5-3(5) ltb and often occur in large numbers in gelatinous colonies among *Sphagnum*. The chloroplast is an axile plate which is frequently obscured by oil droplets. (Id: cell shape, pigmentation, shape of zygospores; 6). Conjugation is fairly frequent.

M. macrococcum is a common form with cells 25-35 µm long,

1.5-2.5 ltb and quadrate zygospores with toothed margins (Plate 42 Fig. 2). *M. violascens* occurs in sunny situations and has a violet pigmentation. The cells are 25-35 µm long, ltb 1.5-2.5 and the chloroplast has a single pyrenoid.

A related genus is *Spirotaenia* which has helical chloroplasts. The species occur on damp rocks in the west but they are all rare. (Plate 41 Figs. 4 and 5).

89. Roya

The cells of *Roya* are more strongly bent and elongate than those of *Mesotaenium* and the species are normally found in wetter sites in the west. The chloroplast is long and ribbon-like, sometimes with a few fine longitudinal ridges and it is narrowed near the cell centre where the nucleus lies. A few species resemble *Closterium* in the possession of apical vacuoles but they differ in their chloroplast structure and cell shape. (Id: cell shape; 3).

R. obtusa (Plate 42 Fig. 1) is sometimes found among mosses on wet rocks or at lake margins. The cells are 75-150 µm long, 7-9 ltb.

90. Cylindrocystis

These species are often associated with *Mesotaenium* in base-poor fens and marshes and at lake margins. The cells are shortly cylindrical with rounded ends and with two large chloroplasts which are either stellate or a little elongate with longitudinal ridges, each with a pyrenoid. (Id: cell shape, colour and ornamentation of the zygospores; 5).

C. brebissonii is a common form (Plate 42 Fig. 3) with cells 35-55 µm long and 2.5-3.5 ltb. The chloroplast is sometimes obscured by oil droplets and the zygospores, which are uncommon, are dark and angular.

PLATE 41

1 *Nautococcus pyriformis*, Southborough, Kent, x 2200, 2 *Apiocystis brauniana*, Malham, Yorks, colony x 200, 3 *Lagerheimia chodati*, x 1000, 4 *Spirotaenia condensata*, x 300, 5 *S. turfosa*, x 400, 6 *Euastropsis richteri*, x 1300, 7 *Tetrastrum staurogeniaeforme*, x 1700, 8 *Arthrodesmus incus*, x 550, 9 Cell division in *Cosmarium botrytis*, 10 *Cosmarium margaritiferum*, in face view, s) semicell; si) sinus; i) isthmus. x 900, 11 *C. margaritiferum*, Side view, 12 *C. margaritiferum*, Apical view.

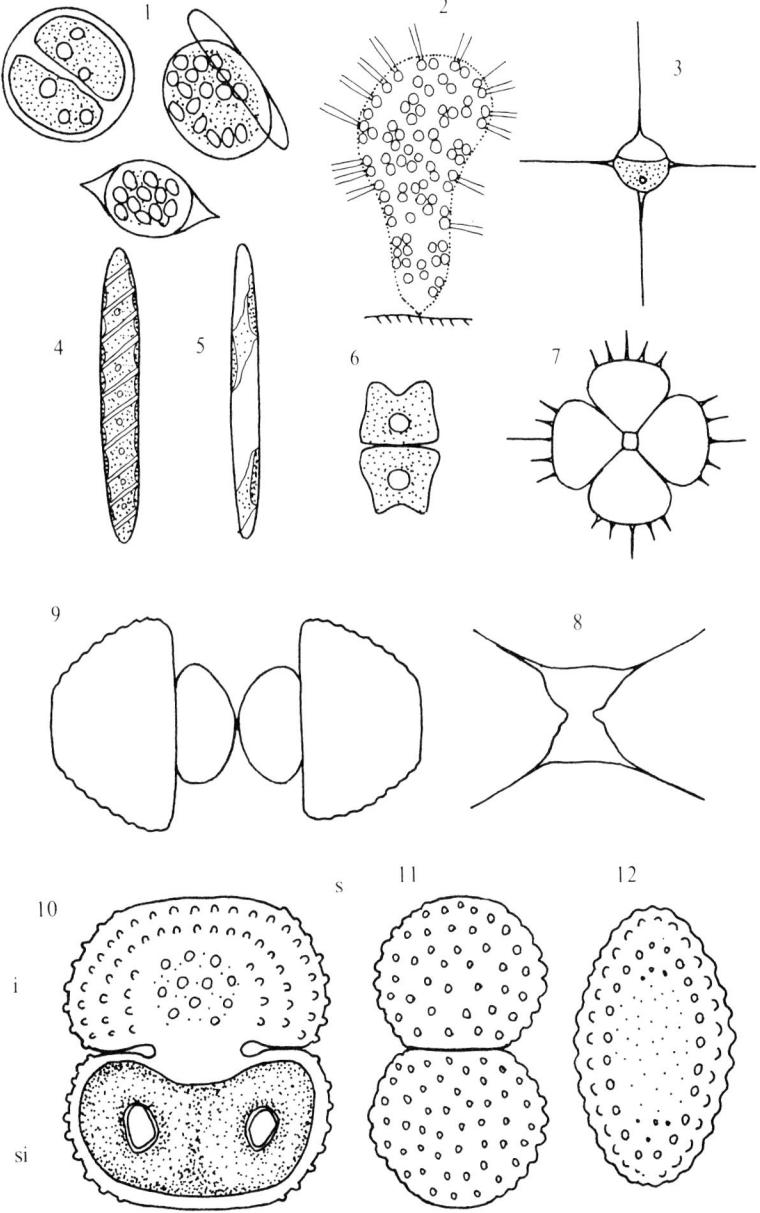

91. Netrium

This is a genus closely related to the last but distinguished by the elongate, deeply ridged chloroplasts which are regularly incised giving the edges a saw-like form. The species are sometimes common among *Sphagnum* in the north and west. (Id: cell shape; 3).

N. digitus is the commonest species with cells 120-400 µm long, 3-4 ltb and about 20-40 µm wide at the apices. The chloroplasts possess several elongate pyrenoids. (Plate 42 Fig. 4).

92. Tetmemorus

The cells of this small genus are very distinctive with an obvious notch at the apices and occasionally a slight median constriction. The cells are cigar-shaped and measure 50-250 µm in length, 3-6 ltb. The chloroplasts are axile and longitudinally ridged with several pyrenoids and the cell walls are often minutely granular. (Id: cell shape, wall ornamentation, notch shape; 5).

T. granulatus (Plate 42 Fig. 9) is often abundant among wet *Sphagnum* with fusiform cells 120-240 µm long, 3.5-5 ltb. The walls are covered with small, evenly distributed granules. Monstruous, bifurcate forms are occasionally seen in this and other species.

93. Penium

Cells cylindrical with rounded, or occasionally tapered ends and often with a slight median constriction. Minute pores can be distinguished in the cell wall under high magnification and these sometimes form regular rows. The chloroplasts are composed of ribbed axile cylinders each with a pyrenoid. In some species a

---

PLATE 42

1 *Roya obtusa*, x 2000, 2 *Mesotaenium macrococcum*, Malham, Yorks, x 1600, 3 *Cylindrocystis brebissonii*, Malham, Yorks, x 1100, 4 *Netrium digitus*, Wastwater, Cumbria, x 1100, 5 *Pleurotaenium trabecula*, Bayham, Sussex, x 1100, 6 *Penium polymorphum*, x 1100, 7 *Euastrum cuneatum*, Ben Cruachan, Argyll, x 1100, 8 *E. ventricosum*, Ben Cruachan, Argyll, x 1100, 9 *Tetmemorus granulatus*, Wastwater, Cumbria, x 1100.

(.75)

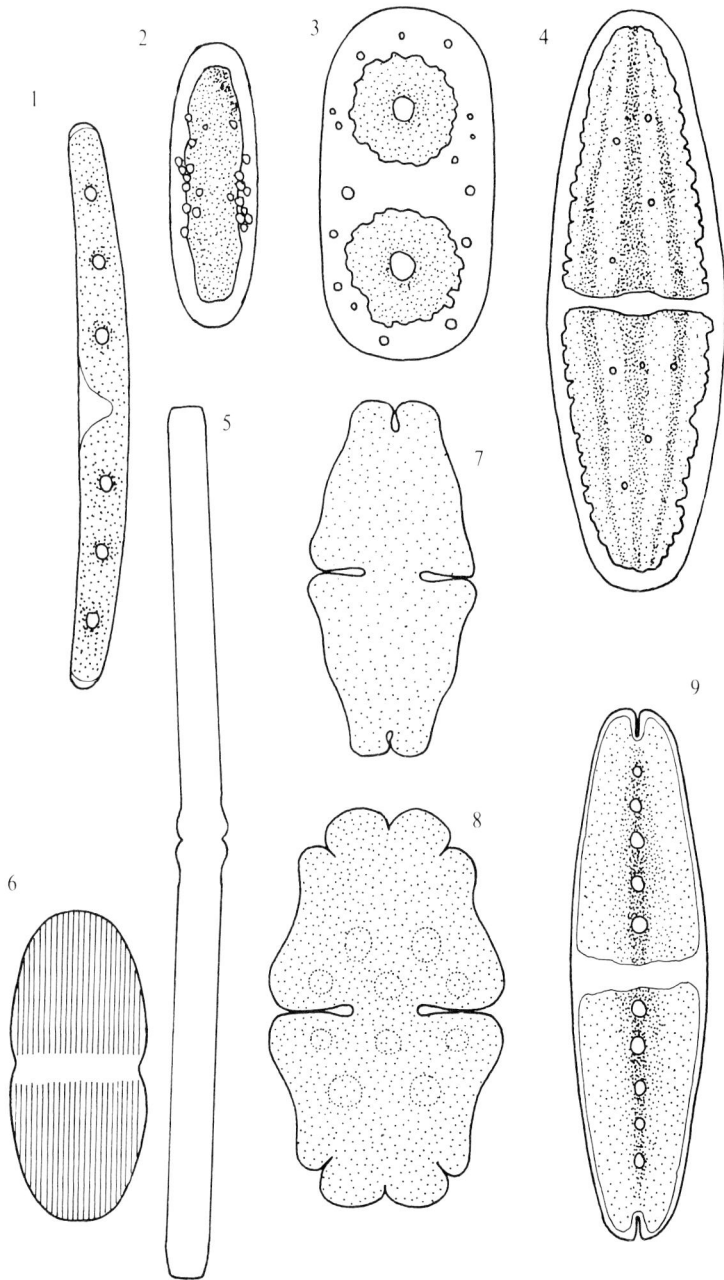

203

clear transverse line or suture can be seen girdling the cell. The cells measure (20)30-100(360) µm in length, (2)3-8(35) ltb. (Id: cell shape, wall ornamentation, presence of a girdle, number of pyrenoids; 100).

The species are found among *Sphagnum*, often at lake margins but they are rarely common. *P. polymorphum* (Plate 42 Fig. 6) is a widespread, montane species with cylindrical cells 40-60 µm long, 1.8-2.5 ltb with a slight median constriction. The wall is finely striated.

94. Closterium

Cells sickle-shaped, bow-shaped or fusiform with finely attenuated apices, measuring (50)100-500(1000) µm in length and 5-25(50) ltb. The two chloroplasts are clearly ridged in the larger forms and contain several pyrenoids which are either scattered or in rows. A small spherical vacuole is usually present at the apex of the chloroplasts and this often contains crystalline inclusions showing Brownian motion. The cells walls are either smooth or finely striated and one or more transverse bands may be present. The walls are also frequently stained brown by iron or humic material. Some authorities quote the dorsal arc length (DA) for the bow-shaped forms. A method for calculating this is given in the appendix. Species of *Closterium* are among the most widespread and abundant desmids and are often seen conjugating. They occur in both eutrophic and oligotrophic waters and appear to be well adapted to living on the surface of fine sediments. In common with *Micrasterias* and a few other forms, they have a limited motility which appears to result from mucilage secretion. Dorsal arc lengths range from (5°)20-130°(200°). (Id: cell shape, form of the apices, wall ornamentation, presence of girdle bands, arc length, shape of zygospores and ornamentation; 100).

PLATE 43

1 *Closterium cornu*, Ashdown Forest, Sussex, x 325, 2 *C. kuetzingii*, x 250, 3 *C. acerosum*, Glen Coe, x 325, 4 *C. juncidum*, cell x 350, striations, x 800, 5 *C. setaceum*, x 250, 6 *C. striolatum*, Ben Cruachan, Argyll, showing part of cell with striations, x 325, 7 *C. littorale*, R. Lochy, Argyll, x 325, 8 *C. parvulum*, Forest Burn, Northumbria, x 375, 9 *C. ehrenbergii*, Speldhurst, Kent, x 200.

(.75)

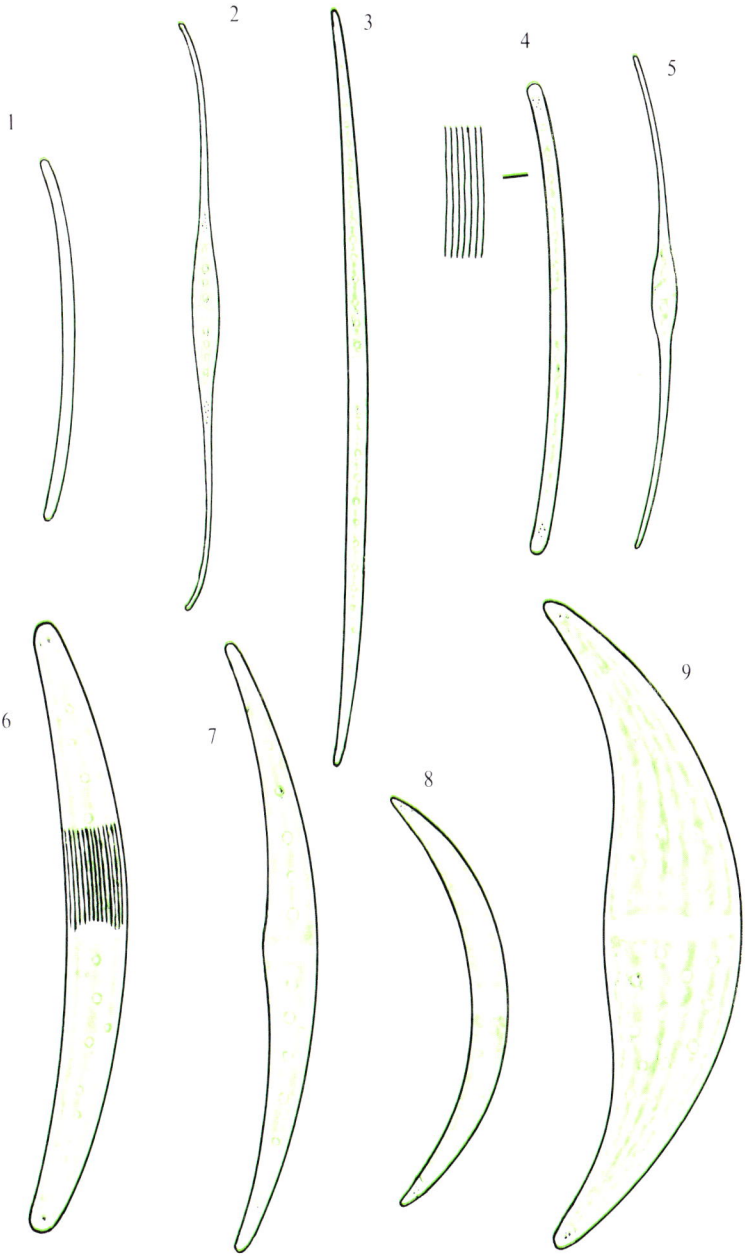

Key to the illustrated species

1a   Cells strongly curved and sickle-shaped. Chloroplast extending to cell apex, DA $> 30°$.   2

1b   Cells narrow and straight or slightly bow-shaped, DA $< 30°$.   5

2a   Ventral margin gibbous. Chloroplast with scattered pyrenoids. Cells 350-550 µm, 3.5-5 ltb, DA 110-135°. Widespread.   *C. ehrenbergii* (Plate 43 Fig. 9)

2b   Ventral margin smooth, pyrenoids usually a single row.   3

3a   DA 110-130°, cells 90-130 µm, 9-12µm, 9-12 ltb. Common and widespread.   *C. parvulum* (Plate 43 Fig. 8)

3b   DA 30-70°.   4

4a   Cells 150-300 µm, 9-12 ltb, DA 35-65°. Walls plain or finely striated, occasional in base poor sites.   *C. littorale* (Plate 43 Fig. 7)

4b   Cells 230-480 µm, 9-12 ltb with broadly rounded apices, DA 30-70°, walls often tinted brownish and finely striated with girdle bands. Occasional.   *C. striolatum* (Plate 43 Fig. 6).

5a   Cells attenuated into long, colourless apices, walls finely striated.   8

5b   Cells not attenuated into long colourless apices, walls smooth or striated.   6

6a   Walls finely striated   7

6b   Walls smooth, with slightly attenuated apices, cells 170-380 µm, 20-30 ltb.   *C. cornu* (Plate 43 Fig. 1)

7a   Apices attenuated, cells 300-460 µm, 10-40 ltb, frequent.   *C. acerosum* (Plate 43 Fig. 3)

7b   Apices rounded, not attenuated, cells 100-330 µm, 25-45 ltb, frequent.   *C. juncidum* (Plate 43 Fig. 4)

8a   Pyrenoids 4-5 per chloroplast, cells 350-530 µm, 20-30 ltb, apices gradulally attenuated.   *C. kuetzingii*
(Plate 43 Fig. 2)

8b   Pyrenoids two per chloroplast, apices fine and untapered, cells 200-450 µm, 25-30 ltb. *C. setaceum* (Plate 43 Fig. 5)

95. Pleurotaenium

Cells elongate-cylindrical with truncated apices, 200-1000 µm long, and 8-24 ltb with a slight, often wavy median constriction. The genus is easily distinguished by the long cells and the semicells which are often slightly out of line with each other. The chloroplasts consist of a long ribbon with or without longitudinal ridges, or of a series of long, narrow discontinuous ribbons with several pyrenoids. Apical vacuoles often occur and the cell apices occasionally have short teeth or spines. (Id: type of chloroplast, cell wall ornamentation, shape of apices; 50).

P. trabecula (Plate 43 Fig. 5) is a common British species which occurs among mosses in lakes and stream throughout the country. The chloroplasts consist of a series of longitudinal bands and the cells measure 350-700 µm in length, 10-22 ltb. The apices are smooth but the wall is minutely granular throughout.

96. Euastrum

The cells of *Euastrum* are flattened in vertical view and tend to lie on their broad faces. The distinctive feature of the cells is their apical notch or indentation and the deep median constriction. The cells range from 20-100(200) µm in length, and 1-2.5(3.5) ltb in face view. In apical view they are about 3-4 times as broad as deep. The cell outline is also interrupted by two or four additional notches each side of the isthmus. Most species have a prominent bulge at the centre of the semicell and the walls may be either smooth or granular. (Id: cell shape, depth of the apical notch, number and shape of marginal lobes, wall ornamentation; 200). The species are commonly collected among aquatic plants, particularly in base-poor districts. The species illustrated are frequent in the north and west.

1a	Apical notch deep and narrow.	2
1b	Apical notch broad and shallow.	3

2a Apex deeply bilobed with a broad lateral lobe. Semicells with five regular bulges, cell wall coarsely punctate. Cells 80-140 µm long, 1.5-1.7 ltb. *E. ventricosum* (Plate 42 Fig. 8)

2b Apical lobe not distinct from a lateral lobe, cell walls finely punctate or roughened, semicells wedge-shaped with a truncated apex and almost straight lateral margins. Cells 90-150 µm long, 1.8-2.2 ltb. *E. cuneatum* (Plate 42 Fig. 7)

3a Semicells with three roughly equal lobes with broadly infolded faces and three bulges. Cell walls coarsely granular, cells 90-100 µm long, 1-1.2 ltb.   *E. verrucosum* (Plate 45 Fig. 4)

3b Semicells with four broad lobes, sometimes developed into faintly angular apices and a single central bulge. Cell walls smooth, cells 15-30 µm long, 1.2-1.4 ltb.   *E. binale* (Plate 45 Fig. 5)

97. Micrasterias

Cells in the form of a thin circular or elliptical disc with many deep and regular incisions. A 'polar' lobe is usually present which differs in shape from the marginal lobes and is often less deeply incised. The cells are large, from 100-450 µm wide and in common with other large forms the genus can be identified in the field with a x10 hand lens. These attractive plants are not often common and are most likely to be found among old, wet *Sphagnum* and *Utricularia* in the west. The cells show considerable variation and intermediate forms are frequently

---

PLATE 44

1 *Micrasterias jenneri*, x 400, 2 *M. thomasiana*, x 220, 3 *M. denticulata*, x 230, 4 *M. americana*, x 400, 5 *M. rotata*, x 250, 6 *Xanthidium armatum*, Robinson's Tarn, Cumbria, x 400, 7 *X. antilopeum*, Wray Mires Tarn, Cumbria, x 550, 8 *Staurastrum teliferum*, Malham, Yorks, x 850.

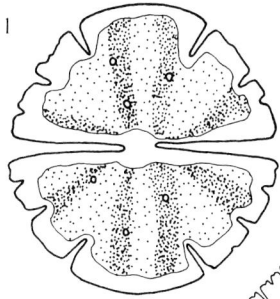

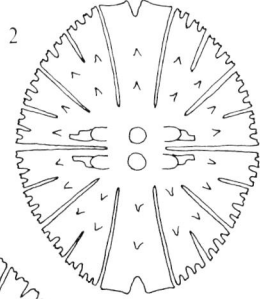

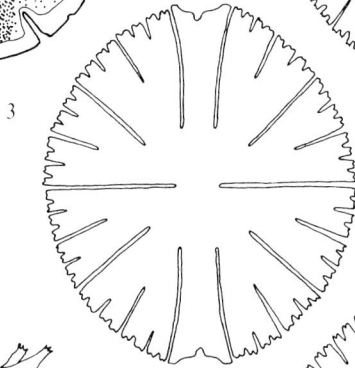

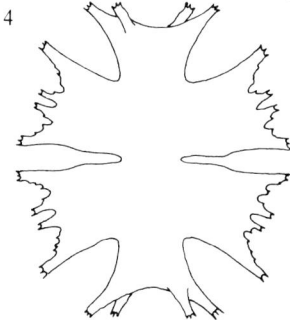

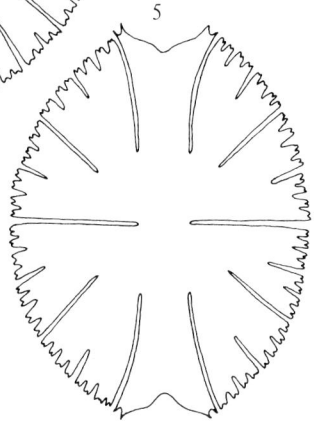

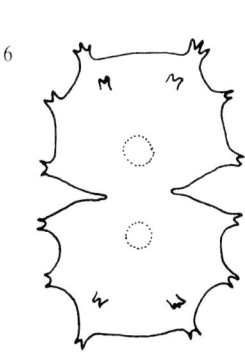

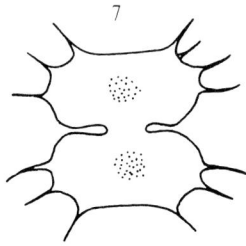

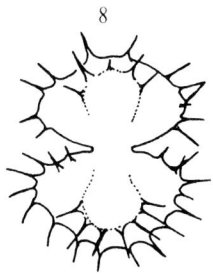

found. (Id: number and depth of incisions, form of the polar lobes, presence of teeth on cell surface, shape of the lobe apices; 50). A key is provided to nine frequent British species.

1a   Polar lobes smooth and linear, without incisions. Lateral lobes with two lateral incisions and forked apices. Cells 50-80 µm.   *M. pinnatifida* (Plate 45 Fig. 1)

1b   Polar lobes with a median incision or indentation.   2

2a   Lateral lobes with numerous narrow incisions.   3

2b   Lateral lobes with about 8 wide incisions.   7

3a   Incisions $< \frac{1}{4}$ of cell width. Apex of polar lobe broad and only slightly indented. Incisions all narrow and straight, cells 120-170 µm.   *M. jenneri* (Plate 44 Fig. 1)

3b   Some incisions $> \frac{1}{4}$ of cell width, polar lobe notched or with spines.   4

4a   Semicells with a pair of teeth on the cell face near the isthmus and with short teeth scattered over the cell surface. Cells 160-220 µm.   *M. thomasiana* (Plate 44 Fig. 2)

4b   Teeth absent from cell face.   5

5a   Apices of lateral lobes truncated, cell wall smooth, cells 180-350 µm.   *M. denticulata* (Plate 44 Fig. 3).

5b   Apices of lateral lobes sharp.   6

6a   Polar lobes with a pair of teeth offset from the cell margin, cells 160-200 µm.   *M. sol* (Plate 45 Fig. 3)

6b   Polar lobes without offset teeth, cells 200-360 µm.   *M. rotata* (Plate 44 Fig. 5)

7a   Polar lobes with large, offset processes. Lateral lobes with unequal and moderate indentations, cells 120-160 µm.   *M. americana* (Plate 44 Fig. 4)

7b   Polar lobes without offset processes. lateral lobe apices narrow and bifid, cells 120-200 µm. *M. crux melitensis* (Plate 45 Fig. 2)

98. Xanthidium

Semicells rounded or quadrate in face view, with a prominent central bulge in apical view and united by a narrow isthmus. The spines are usually narrow and sharp, often in pairs and radiating outwards, 4-10(20) per semicell. The wall may be smooth or finely granular and in a few species the spines are replaced by sharp teeth. There are 1-several chloroplasts each semicell and the cells (10)20-80(130) µm long, 0.9-1.5 ltb, excluding spines. (Id: type and number of spines, shape of semicell, ornamentation of cell wall; 80). The species are most common in mountain lakes and tarns, in the plankton or among aquatic vegetation.

X. armatum (Plate 44 Fig. 6) is easily recognized by the groups of 3-4 tooth-like spines regularly arranged so that 6-8 groups at the semicell margin. Cells 80-130 µm long, 1.4-1.6 ltb. *X. antilopeum* (Plate 44 Fig. 7) is a common species with two pairs of long, diverging spines on each side of the semicell, cells 40-80 µm long, diverging spines on each side of the semicell, cells 40-80 µm long, 1-1.2 ltb, excluding spines. *X. cristatum* is similar but it has further isolated spines close to the isthmus and more rounded semicells (Plate 45 Fig. 6). Cells 40-55 µm long, 1-1.2 ltb.

99. Staurastrum

Cells angular in apical view, the angles numbering 3-10, but three in most of the species and these are either thickened at the corners or drawn out into teeth or arm-like spines. The cells measure (10)30-60(140) µm long, and 0.4-2 ltb. There is an axile chloroplast in each semicell. This is a large genus which can be subdivided into two major groups, those species without corner thickenings or arms (subgenus *Prostaurastrum*) and those with them (subgenus *Staurastrum*). (Id: number of angles in apical view, distribution and type of ornamentation, spines, arms, teeth, punctae, depth of the median constriction; 1200). The species are widespread but more abundant in montane districts and usually occur among aquatic vegetation. A few are truly planktonic and occur in both oligotrophic and eutrophic waters.

A closely related genus is *Arthrodesmus* where the semicells are oval and two angled in apical view with spines at the corners (Plate 41 Fig. 8).

Key to the illustrated species

1a   Angles of semicells with or without sharp spines or teeth, not developed into arms.   2

1b   Angles of semicells developed into broad arms.   7

2a   Spines and teeth absent from the angles, cell walls granular.   3

2b   Spines or teeth present at the angles, common in bog pools.   4

3a   Sinus narrow, walls finely punctate, cells 25-40 μm, 1.1-1.4 ltb.   *S. orbiculare* (Plate 45 Fig. 7)

3b   Sinus wide, walls coarsely punctate, cells 25-40 μm, 1-1.2 ltb.   *S. punctulatum* (Plate 46 Fig. 1)

4a   Spines clustered at angles and scattered elsewhere, cells 30-60 μm, 1-1.3 ltb, widespread.   *S. teliferum* (Plate 44 Fig. 8)

4b   Spines present only at the angles.   5

5a   Angles with several small teeth, cell walls granular only at angles which are rounded. Cells 25-35 μm, 0.9-1.1 ltb.   *S. denticulatum* (Plate 46 Fig. 2)

5b   Angles with one smooth spine, sinus open.   6

6a   Spines pincer-like, semicells oval, cells 25-45 μm, 0.8-1.1 ltb. Common in bogs.   *S. dickei* (Plate 45 Fig. 8)

6b   Spines roughly parallel, semicells triangular in face view, cells 35-50 μm, 0.7-1 ltb, often planktonic.   *S. megacanthum* (Plate 46 Fig. 3)

PLATE 45

1 *Micrasterias pinnatifida*, Robinson's Tarn, Cumbria, x 580, 2 *M. crux melitensis*, Robinson's Tarn, Cumbria, x 400, 3 *M. sol*, Robinson's Tarn, Cumbria, x 300, 4 *Euastrum verrucosum*, Loch Achtriochtum, Glen Coe, x 500, 5 *E. binale*, Eridge, Sussex, x 1400, 6 *Xanthidium cristatum*, Loch Avich, Argyll, x 550, 7 *Staurastrum orbiculare*, R. Lochy, Argyll, x 1200, 8 *S. dickei*, Malham, Yorks, x 800, 9 *Cosmarium ralfsii*, Loch Avich, Argyll, x 380.

(.76)

7a Arms in two whorls, 9 in lower, 6 in upper whorl with the apex of each arm bifid or trifid, arms granular. Cells 60-160 µm, 1.1-1.7 ltb. Handsome species occasional in bogs. *S. arctison* (Plate 46 Fig. 4).

7b Arms in one whorl, 3-4, trifid at apices, common planktonic species. 8

8a Cells with a well defined ring of granules on either side of the isthmus, cells 35-60 µm, 0.4-0.8 ltb. *S. cingulum* (Plate 46 Fig. 5).

8b Cells lacking such a ring, 40-60 µm, 0.4-0.7 ltb.
    S. planktonicum (Plate 46 Fig. 6).

100. Cosmarium

Cells dumb-bell shaped with a narrow or wide, shallow or deep median constiction, with or without ornamentation, measuring (10)20-110(200) µm long, (0.8)1-2.4(2.6) ltb, sometimes with a slight central bulge evident in apical view. The semicells sometimes have a truncated or polygonal appearance in face view but they are not notched as in *Euastrum*. Spines are absent but the walls are often entirely or partly covered in small or large granules in regular or irregular rows. There are 1-several chloroplasts per semicell with 1-several pyrenoids. (Id: cell shape, presence and type of ornamentation pattern of ornamentation, cell outline, shape in apical view, number of chloroplasts and pyrenoids, ornamentation of zygospores; 2000). The species are widespread but more common in montane districts and usually found among aquatic vegetation. Only a handful of the common and distinctive species are illustrated.

PLATE 46

1 *Staurastrum punctulatum*, Pevensey, Sussex, x 1800, 2 *S. denticulatum*, R. Lochy, Argyll, x 900, 3 *S. megacanthum*, Malham, Yorks, x 1300, 4 *S. arctison*, Loch Avich, Argyll, x 900, 5 *S. cingulum*, Windermere, Cumbria. x 600, 6 *S. planktonicum*, Loch Avich, Argyll, x 680, 7 *Cosmarium melanosporum*, Buttermere, Cumbria, x 1800, 8 *C. circulare*, Glen Coe, x 1100, 9 *C. portianum*, Malham, Yorks, x 750, 10 *C. cucurbita*, Ben Cruachan, Argyll, x 1700.

(.75)

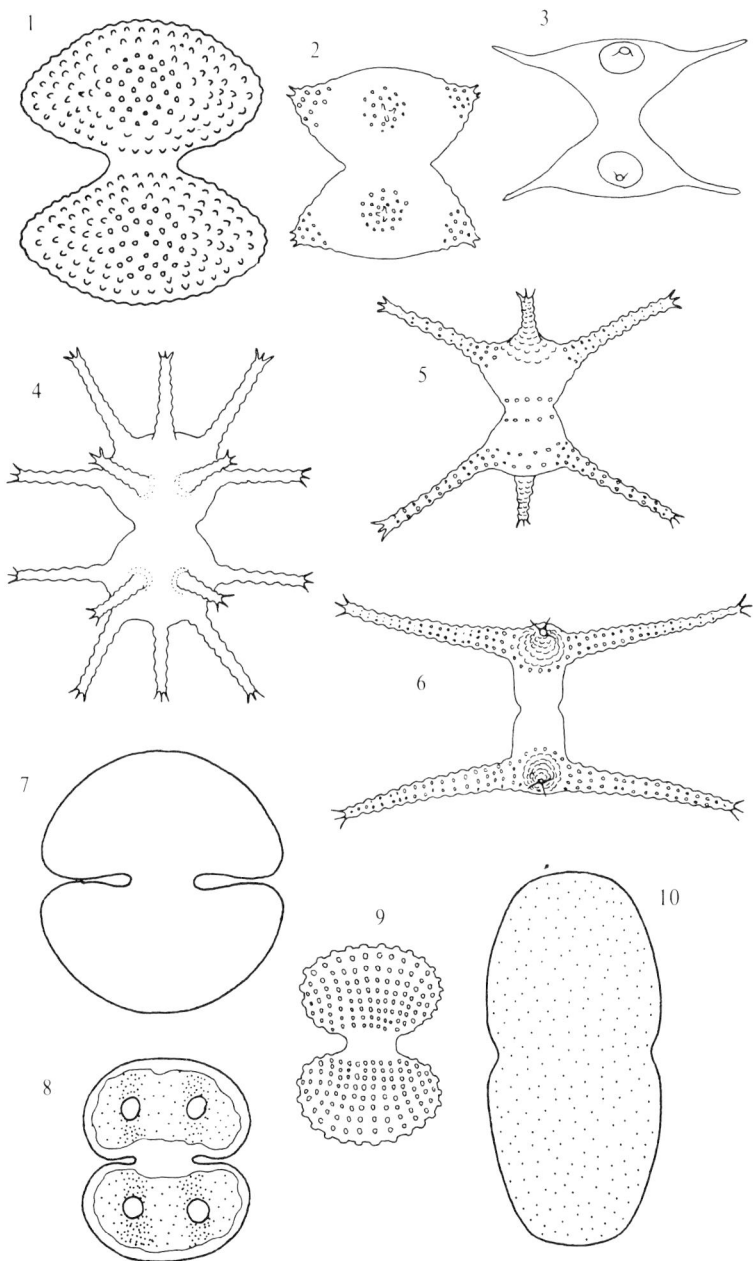

215

Key to illustrated species

1a   Walls smooth or minutely granular.   2

1b   Walls coarsely granular.   5

2a   Median constriction slight, sinus open, semicells rounded-conical with flattened apex, cells 30-50 µm, 2-2.2 ltb.
C. cucurbita (Plate 46 Fig. 10)

2b   Median constriction deep, sinus narrow.   3

3a   Semicells subsemicircular or semielliptic.   4

3b   Semicells broadly conical, wall finely punctate, cells 110-120 µm, 1.1-1.3 ltb.   *C. ralfsii* (Plate 45 Fig. 9)

4a   Cells 15-20 µm, 1-1.2 ltb, semicells semielliptic, smooth.
C. melanosporum (Plate 47 Fig. 7)

4b   Cells 30-95 µm, 1-1.1 ltb, semicells subsemicircular, walls minutely punctate.   *C. circulare* (Plate 46 Fig. 8)

5a   Sinus wide and open, semicells elliptical with about 10 vertical rows of granules, cells 30-40 µm, 1.2-1.4 ltb.
C. portianum (Plate 46 Fig. 9)

5b   Sinus narrow and closed   6

6a   Semicells with a crenate margin, subsemicircular, cells 20-40 µm, 1-1.2 ltb.   *C. subcrenatum* (Plate 47 Fig. 3)

6b   Semicells with a smooth margin.   7

7a   Semicells kidney-shaped with about 30 granules showing on the margin. Cells 60-105 µm, 1.2-1.4 ltb.
C. margaritatum (Plate 47 Fig. 1).

7b   Semicells semielliptic with coarse granules unevenly distributed over the surface, not appearing on the apex margin. Cells 47-55 µm, 1.1-1.2 ltb.   *C. praemorsum* (Plate 47 Fig. 2)

Four further species with smooth cell walls are illustrated in Plate 46.

101. Cosmocladium

Cells dumb-bell shaped, resembling *Cosmarium* but forming gelatinous planktonic colonies with cells connected to each other by fine threads and measuring 10-30 µm long, 1-2 ltb. The species are occasionally abundant in small oligotrophc, montane lakes. (Id: cell shape, form of the colony; 6).

C. saxonicum (Plate 47 Fig. 8) is one of the larger species with cells 22-27 µm long and 1.5-1.7 ltb with deeply constricted cells. The colonies contain a few to several hundred cells which are sometimes found in short parallel rows.

102. Oocardium

The only species, *O. stratum* forms small greenish calcified nodules up to 5 mm across in calcareous streams. The cells are densely clustered together with individual mucilage tubes developing from the base. The median constriction is slight and the cells are often slightly asymmetrical, measuring 13-20 µm long and 1.4-1.7 ltb. This species forms extensive tufa deposits on the continent but is rare in Britain, being restricted to a few streams flowing over the Carboniferous limestone (Plate 47 Fig. 9).

103. Gonatozygon

Cells elongate-cylindrical, 70-400 µm in length and 10-20(40) ltb. The walls are either smooth or punctate and granular and the cells are often found in short ribbons. Each cell has a long ribbon-like chloroplast with a series of pyrenoids. The species are frequent in oligotrophic lakes and tarns. (Id: form of chloroplast, number of pyrenoids, wall ornamentation; 6).

G. brebissonii is a frequent form with long, narrow cells and slightly capitate apices (Plate 47 Fig. 10). The chloroplasts have 5-16 pyrenoids and the wall is minutely granular or smooth, 160-290 µm long, 20-35 ltb.

A related genus is *Genicularia* which differs in having helical chloroplasts. The species are very rare in Britain.

104. Desmidium

Cells with strongly lobed or angled margins when seen in vertical view and generally shorter than wide in face view, 0.3-1 ltb, 12-50 µm wide. The long filaments often have a twisted appearance and each cell has a small, notch-like median constriction. In a few species, the constriction is absent. These desmids are occasionally common among vegetation at the margins of montane, oligotrophic lakes. (Id: number and form of lobes, depth of constriction, development of lens-shaped space between cells, shape of zygospores; 20).

D. swartzii (Plate 38 Fig. 2) is a species with triangular cells in apical view which measure 37-50 µm wide, 0.3-0.5 ltb. The gap between the cells is narrow or absent and the angles are developed into four rounded nodules which form a helical trace down the filament.

105. Spondylosium

Cells in vertical view elliptic, in face view with a deep, fairly narrow median constriction. The cells measure (5)8-25 µm long, 0.7-1 ltb with a chloroplast in each semicell. (Id: shape of cells in face view; 20).

S. planum (Plate 38 Fig. 1) is a small species frequently recorded with smoothly rounded and deeply constricted cells 9-20 µm wide and 0.7-0.9 ltb. The species occur at lake edges in mountainous districts where they are sometimes frequent. A related genus, *Sphaerozosma*, is distinguished by the presence of two small peg-like connecting processes which occur between adjacent cells.

PLATE 47

1 *Cosmarium margaritatum*, Malham, Yorks, x 750, 2 *C. praemorsum*, x 700, 3 *C. subcrenatum*, x 1350, 4 *C. contractum*, semicell, x 1200, 5 *C. granatum*, semicell, x 1100, 6 *C. moniforme*, semicell, x 1300, 7 *C. impressum*, semicell, x 1600, 8 *Cosmocladium saxonicum*, Robinson's Tarn, Cumbria, showing bacteria in the mucilage, x 500, 9 *Oocardium stratum*, Malham, Yorks, calcified colonies xl, cells x 750, 10 *Gonatozygon brebissonii*, cell x 400, detail of apex x 1500, 11 *Hyalotheca mucosa*, Loch Ba, Glen Coe, x 750, 12 *Bambusina brebissonii*, Intake Tarn, Cumbria, x 750.

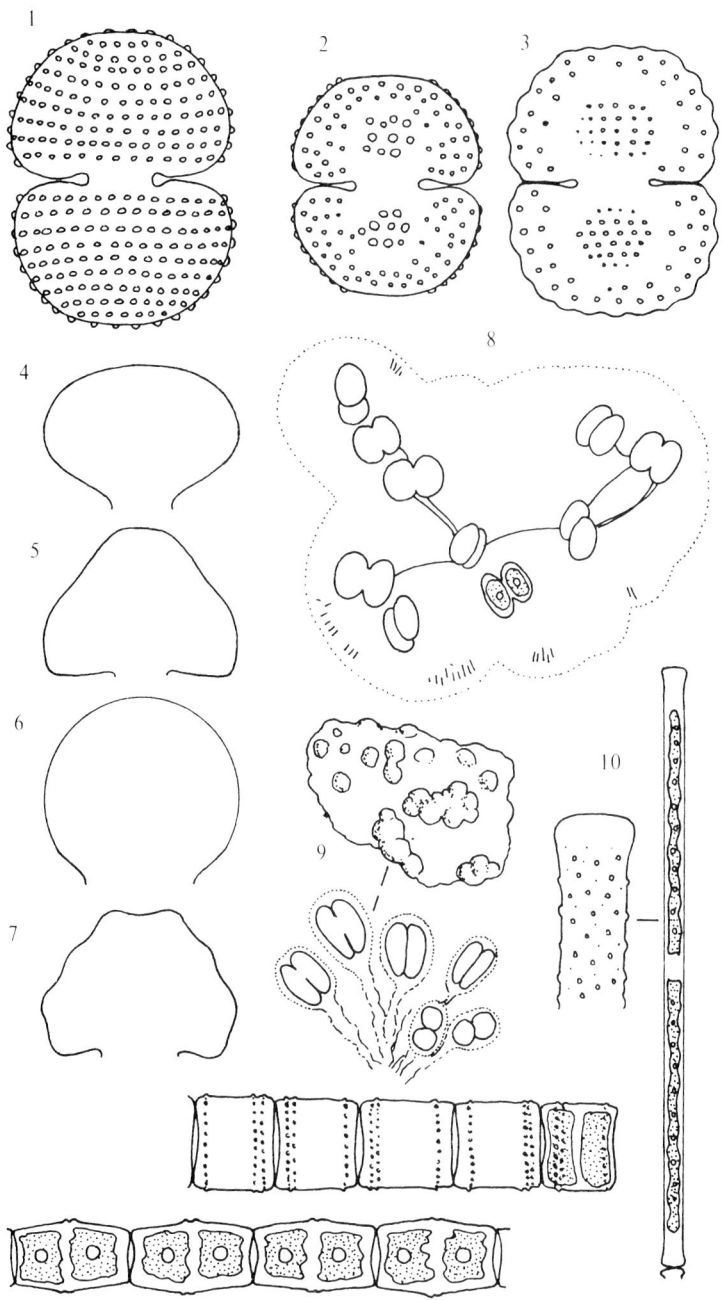

106. Bambusina

Cells barrel-shaped with a slight notch in the mid-region and usually a transverse series of minute pores. Vertical view circular. Each semicell has an axile chloroplast with a central pyrenoid. (Id: cell shape; 10).

B. brebissonii is occasionally seen among other filamentous desmids with cells 25-35 µm long, 0.5-0.7 ltb. The cells may have fine vertical striations near the apices (Plate 47 Fig. 12).

107. Hyalotheca

Cells shortly cylindrical with a broad and extremely shallow median constriction which may be altogether absent in a few species. In some forms, transverse rows of pores occur as in *Bambusina*. Each cell has two stellate chloroplasts with a central pyrenoid and conjugation has been observed frequently. The species are occasionally found in similar habitats to *Bambusina* in montane, oligotrophic sites. (Id: cell shape development of constriction, shape of zygospores, presence of pores; 12).

H. mucosa (Plate 47 Fig. 11) is frequently recorded. The cells are 15-20 µm wide and 0.7-0.9 ltb, with no median constriction. Two transverse rows of pores occur each side of the cell and the filaments, in common with many other desmids, are invested in a broad, hyaline, mucilage envelope.

PLATE 48

Figs. 1-5 *Chara vulgaris*

1 Habit x1 from Whin Park, Orkney, 2 Fertile lateral with 3 bracts, antheridium and oogonium x 60, 3 Section throuth main axis showing the distichous cortex and internode cell. x 100, 4 Sperm x 500, 5 Detail of whorl showing cortex, spines, stipulodes and laterals x 150, 6 *Chara hispida* habit x 1 from Whittington, Shropshire, 7 *C. hispida*, detail of cortex with pairs of spines x 15, 8 *Nitella flexilis* habit x 1 from Allan Tarn, Cumbria, 9 *N. flexilis* detail showing a dectyl. x 10, 10 *N. translucens* from Burghfield, Bucks, whorl showing short, 2-celled dactyls x 10, 11 *N. translucens* oogonium x 90.

a antheridium; br bract; c cortex; co corona; d dactyl; et enveloping thread; in internode; l lateral; n node; oo oogonium; r rhizoids; s spine; st stipulode.

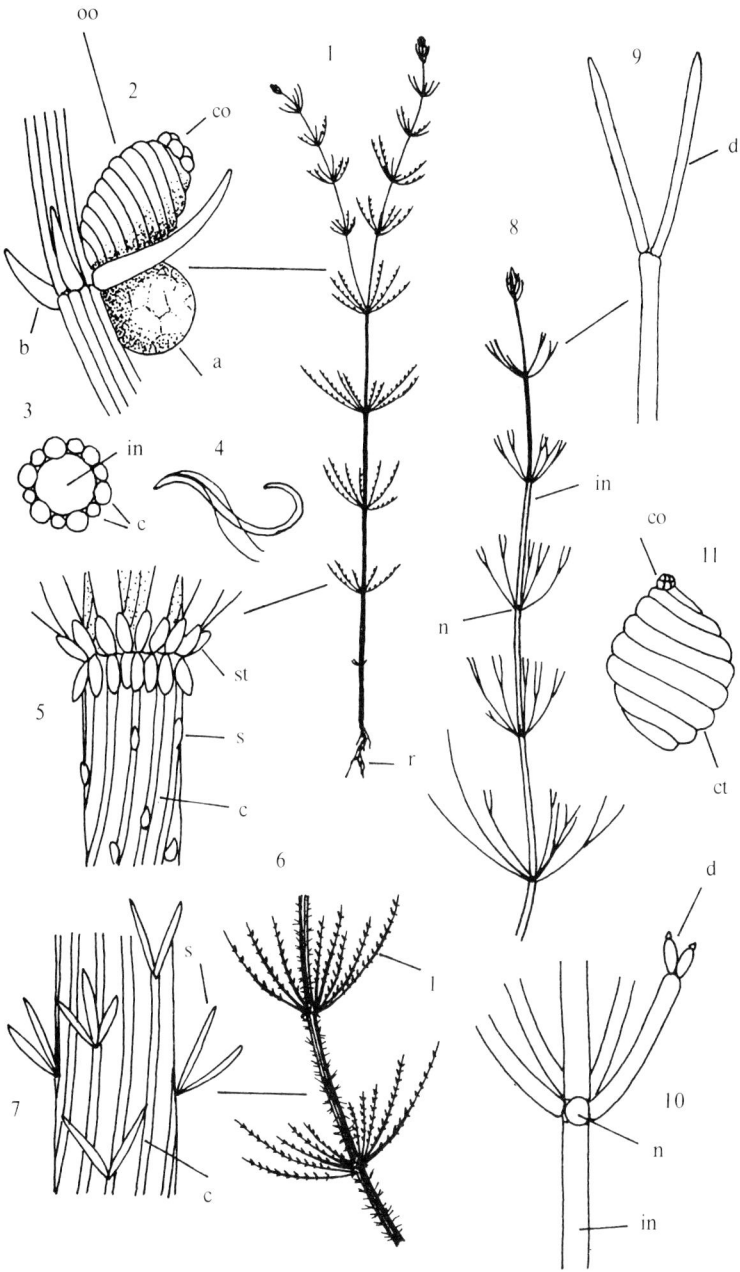

THE CHAROPHYCEAE

This class contains a small number of genera with a high degree of complexity and specialization. The plants are always attached and often grow in dense swards to a height of 50 cm or more. The cells are usually grass-green in colour but species of *Chara* may become densely encrusted in calcium carbonate and then appear almost white. This gives the plants a brittle texture and accounts for their common name of 'stoneworts'. Fresh specimens of *Chara* have a smell similar to sweetened garlic which some people find unpleasant and others attractive.

The main axis of all charophytes is composed of two types of cells, long *internode* cells (Plate 48 Fig. 8,10) up to 10 cm in length, alternating with much shorter *node* cells from which whorls of laterals arise. The main axis is attached to the soft substratum by numerous branched rhizoids and the axis itself may be branched several times (Plate 48 Fig. 1). In *Chara* and a few related genera, the internodes are clothed in a layer of narrow cortical cells which develop from the nodes (Plate 48 Fig. 3,5,7). They normally have a slight helical twist. Some of the cortical cells develop small outgrowths called *spines* (Plate 48 Fig. 5,7) whilst the whorls of some genera are ringed below by a series of elongate cells called *stipulodes* (Plate 48 Fig. 5).

Although vegetative growth from the node cells is often prolific, oogamous sexual reproduction is commonly seen. The sex organs are complex and develop as regular sessile offshoots together with several sterile outgrowths or *bracts* on the whorls of laterals (Plate 48 Fig. 2). The conspicuous antheridia are spherical and bright orange or yellow when mature. Large numbers of sperms are released from filamentous cells within these bodies. The sperms have two backward-pointing flagella (Plate 48 Fig. 4) and the cells are narrow and twisted, similar to those of *Coleochaete* (Chlorophyceae).

The female organs, or oogonia are unlike any other in the plant kingdom. The female cell is surrounded by a series of helically twisted threads capped at their apex by one or two rings of small *corona* cells (Plate 48 Fig. 2,11). This ring of cells is important in classification. After fertilization, the oospore wall darkens and the surrounding threads become calcified and silicified. The oospores normally become detached and later germinate to produce a green filamentous protonema similar to that found in the bryophytes. The calcified oospore threads fossilize easily and remains of charophytes have been found in rocks 400 million years old.

Charophytes are large algae and best preserved in a solution of either 5% formalin or 60% alcohol, 30% water and 10% glycerine. They can also be floated onto cartridge paper in the same way as seaweeds. They are widely distributed throughout the world but studied by few people. Today, few graduate biologists have even heard of them.

Key to common genera

Corona of 5 cells, cortex usually present, with a ring of stipulodes below the whorls. Plants often calcified.   *Chara* 1

Corona of 10 cells, cortex and stipulodes absent, rarely heavily calcified.   *Nitella* 2

1. *Chara*

Plants 5-40 (130) cm long, forming dense carpets or straggling groups in shallow lakes, ponds and marshes, usually in hard water. There are about 20 species divided into two sub-genera distinguished by the number of rows of stipulodes below the whorls (1 or 2). Species are distinguished by the type of cortication, form and position of the spines and the size of the oogonia.

The type of cortication is an important character. There is a primary and secondary cortex present in most species and the spines are found only on the primary cortex. There are three types of cortication, a) primary cortex only, spines on all cortex cells (Haplostichous, few spp.), b) spines on alternating cells (Diplostichous) and c) spines on every third row of cells (Triplostichous). In *Chara*, the oogonium is always positioned above the antheridium but in the less common *Lamprothamnium*, the oogonium is below the antheridium.

Two common species of *Chara* are illustrated, both of these have two rows of stipulodes and a diplostichous cortex. *C. vulgaris* (Plate 48 Fig. 1-5) is a variable species, particularly in the length of the internodes. The main axis are 0.5-2 mm wide near the base and the plants 5-30 cm long. The single spines are small and inconspicuous and the oogonia about 0.5 mm long. *C. hispida* (Plate 48 Fig. 6,7) is a more robust plant 20-40 cm in length. The sharp spines develop in groups of 2 or 3 and are conspicuous without a lens and the oogonia about 1 mm long.

2. *Nitella*

The species are simpler in structure than *Chara*, with no cortex or stipulodes below the laterals, which usually number six per whorl. The plants are rarely calcified and most often found in soft water lakes and peaty pools to depths of 10 m or more. About 50 species have been described, distinguished mainly on the structure of the whorls of laterals. The species are less frequently found fertile than *Chara* and appear to 'fruit' earlier in the year, often from May-June compared to July-August for *Chara* (in Britain).

The whorls of laterals are either single or multiple and in the latter, the laterals may be of different lengths. The laterals themselves usually branch once. The terminal branches are termed *dactyls* which consist of one or more cells. The internode cells of *Nitella* can be called 'giant cells' with some justification and they have been widely used, together with those of *Chara* in physiological research. The numerous discoidal chloroplasts can be clearly seen in *Nitella*. The antheridia occur below the oogonia and the latter are elliptical in section. In the related but less common genus *Tolypella*, the oogonia are circular in section and surrounded by a ring of antheridia. The sex organs occur in dense clumps giving a distinctive 'nested' appearance.

Nitella flexilis is a frequent species, 10-40 cm long with single whorls and long, V-shaped unicellular dactyls (Plate 48, Fig. 8,9). *N. translucens* is our commonest species with a similar habit but distinctive, short, 2-celled dactyls (Plate 48 Fig. 10,11).

APPENDIX A

Equation 1. To estimate the number of cells in a *Volvox* colony (N).

$$N = \frac{15\ r^2}{a^2}$$

r = colony radius

a = average distance between adjacent cells.

Equation 2. To estimate the angle of arc (DA) of a curved cell such as *Closterium.*

$$DA = 2\ \tan^{-1}\left[\frac{\dfrac{x^2 + y^2}{2y} - y}{x}\right]$$

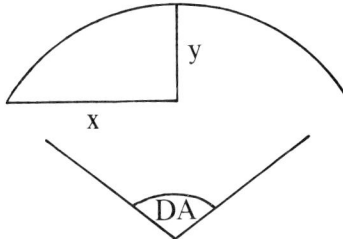

Supplier's addresses

Plankton nets and other sampling equipment.	S.M. Davis, 45 Quest Hills Road, Malvern, Worcs. WR14 1RL.
Light microscopes.	Hampshire Micro, 57 New Market Square, Basingstoke, Hants. RG21 1HW.
Microscope graticules.	Graticules Ltd., Morley Road, Tonbridge, Kent. TN9 1RW.

Societies in Britain and the United States of interest to amateur and professional algologists.

British Phycological Society. Secretary: Dr. J.M. Jones, Department of Marine Biology, Port Erin, Isle of Man.

Queckett Microscopical Club. c/o The British Museum (Natural History), Cromwell Road, South Kensington, London SW7 5BD.

Phycological Society of America. Membership Office, P.O. Box 368, Lawrence, Kansas 66044.

International Phycological Society. Secretary: R.B. Searles, Department of Botany, Duke University, Durham, N. Carolina 27766.

Selected Bibliography

1. *Identification of algae.*

Bourrelly, P. (1966-70) *Les Algues d 'eau Douce* 3 vol., N. Boubee et Cie, Paris. (Comprehensive to genera, good illustrations, expensive).

George, E.A. (1976) A guide to algal keys. *British Phycological Journal* vol. II pp. 49-55. (All important literature cited to this date).

Pascher, A. (1913-36) *Die Susswasserflora Deutschlands, Oesterreichs und der Schweiz* 15 vol., Gustav Fischer, Jena. (Good but outdated keys in German to most genera, obtainable as an expensive reprint from Koeltz).

Patrick, R. & Reimer, C.W. (1966, 1975) *The Diatoms in the United States* 2 vol., Monographs of the Academy of Natural Sciences, Philàdelphia. (Excellent work but incomplete and fairly expensive).

Prescott, G.W. (1951) *Algae of the Western Great Lakes Area* Wm. C. Brown, Dubuque, Iowa. Reprint edition (1982) by Otto Koeltz Science Publishers, P.O. Box 1380, D-6240 Koenigstein, F.R.G. (fairly comprehensive, moderately expensive).

Prescott, G.W. (1978) *How to know the Freshwater Algae* 3rd Edn., W.C. Brown & Co., Iowa. (Useful key to genera, reasonable price).

Smith, G.M. (1950). *Freshwater Algae of the United States* 2nd. Edn., McGraw Hill, New York & London. (Well bound, moderately expensive).

Starmack, C. (1964-   ) *Flora Slodkowodna Polski* 15 vol., Polska Akademia Nauk, Warszawa. (The most comprehensive work on f.w. algae, well illustrated but mostly in Polish, difficult to obtain).

Vinyard, W.C. (1979) *Diatoms of North America*, Richmond Publishing Company.

West, G.S. & Fritsch, F.E. (1927) *A Treatise on the British Freshwater Algae*. Reprint by J. Cramer (1968) obtained from Wheldon & Wesley Ltd., Codicote, Herts. (Much outdated but full of useful information).

Wood, R.D. & Imahori, K. (1964) *A revision of the Characeae* 2 vols., Cramer. (Standard work but expensive).

2. *Ecology and Morphology.*

Fritsch, F.E. (1935) *Structure and Reproduction of the Algae* 2 vol., Cambridge University Press. (Outdated but packed with information, expensive).

Round, F.E. (1981) *The Ecology of Algae*, Cambridge University Press. (Marine & f.w., comprehensive).

3. *Biology*

Brook, A.J. (1981) *The Biology of Desmids*, Blackwell, London. 286 pp.

Carr, N.G. & Whitton, B.A. (Eds.) (1982) *The Biology of Cyanobacteria* Blackwell, London. 688 pp.

Lee, R.E. (1980) *Phycology*, Cambridge University Press, 498 pp. (Useful introduction, well priced).

Werner, D. (Ed.) (1977) *The Biology of Diatoms*, Blackwell, London. 498 pp.

GLOSSARY

acicular	needle-shaped
alveola	pit
apical (cell)	narrowed end bearing the flagella in Chlorophyceae
aplanospore	small, non-motile spore.
areolate	broken up into a mosaic.
attenuated	becoming narrowed to a point.
autospore	aplanospore similar in shape to parent cell or colony.
auxospore	diatom spore produced under adverse conditions and usually associated with sexual reproduction.
axial area (diatom)	clear area of the valve usually containing the raphe.
axile	in a central position.
baeocyte	minute spore produced by internal cleavage in cyanobacteria.
basal cell	specialized cell attaching plant to substratum.
benthic	on the surface of sediment.
bifid	two-pronged.
biseriate	a double row of cells.
bloom	dense growth of algae discolouring the water.
botryoidal	in grape-like bunches.
capitate	swollen at the apices.
carinal dots	row of pores on the keel of *Nitszchia*.
carpogonium	female sex organ of Rhodophyceae.
carpospore	non-motile spore formed from direct or indirect division of zygote in Rhodophyceae.
chloroplast	body containing the photosynthetic pigments.
clavate	club-shaped.
cruciate	cross-shaped.

conjugation	form of sexual reproduction where neither gamete is flagellate.
contractile vacuole	minute, pulsating spherical body, usually situated apically in flagellates.
corticate	covered with 1 or more layers of cells.
costa	rib or thickened bar.
crenate	crinkled.
cuneate	wedge-shaped.
cyst	thick-walled, often silicified resting spore.
cytoplasm	protoplasm of a cell, excluding the nucleus.
daughter cell	cell formed by subdivision of a mother cell.
diffluent	readily disintegrating.
diploid	number of chromosomes in the zygote.
divergent	directed away from the cell.
dendroid	tree-like branching.
dorsal	more convex margin of a dorsiventral cell.
dorsiventral	cell with two margins of different curvature.
dystrophic water	water containing high levels of humic material which may reduce the oxygen content.
epicone	Forward-directed end of cell in swimming dinophyceae.
epilithic	growing upon rocks.
epiphytic	growing upon plants.
eutrophic	nutrient enriched water which may become deoxygenated.
exospore	spore produced by budding in cyanobacteria.
face view (desmid)	View of cell in its most stable position with the median contriction visible.
false branch	branch formed by trichome rupture within the sheath of cyanobacteria.
filiform	thread-like.
fission	simple cell division.
flagellum	whip-like structure used as an organ of propulsion.
frustule	silica shell of a diatom.
fusiform	elongate and tapered towards the ends.
gamete	sex cell.
gametangium	cells containing the gametes.

gas vacuole	protein covered gas packet found in many cyanobacteria.
gibbous	locally swollen.
haploid	cell containing the basic number of chromosomes.
heterocyst	thick walled cell with pores at one or both ends found in cyanobacteria and involved in nitrogen fixation.
heterotrichous	differentiated into a basal and erect system of filaments.
hormogonium	motile trichome fragment serving as a reproductive unit in the cyanobacteria.
hyaline	without colour.
hypocone	rear end of the cell in swimming Dinophyceae.
imbricate	overlapping.
intercalary	cells within the body of a filament.
isogamous	with identical gametes
isopolar	diatom frustule with apices the same size and shape.
isthmus	narrowed mid-region of a desmid.
keel	a prominent ridge.
laminate	layered.
lanceolate	wide in the middle and tapered to a pointed apex.
lax	loosely arranged.
lenticular	lens-shaped.
loculus	chamber.
lorica	urn- or pot-shaped structure enclosing a cell.
lunate	shape of the moon in its first quarter.
median position (cell)	girdled in a central position.
metabolic	exhibiting elastic movements.
mesotrophic	moderately enriched with nutrients.
montane	mountainous country.
mother cell	cell subdividing to produce daughter cells, normally diploid to haploid.
muciferous body	small fusiform bodies often in helical rows and associated with the pellicle of

Euglenoids, staining in Neutral Red dye.

multiseriate	in many rows.
naked cell	cell without an external wall.
nannoplankton	algae which pass through the plankton net.
neustonic	living at the air-water interface.
oligotrophic	nutrient poor water.
oogamous	with an enlarged, non-motile female gamete.
oospore	fertilized oosphere.
papilla	small protrusion of cell wall between the flagella.
paramylum	starch-like storage material produced by Euglenoids.
parietal	close to the cell wall.
pedicellate	borne upon 1 or more support cells.
pellicle	series of fine interlocking protein bands developing beneath the cell membrane of Euglenoids, with a helical arrangement.
periplast	strengthened cell membrane allowing metabolic movements.
posterior (cell)	end of cell opposite to that bearing flagella in Chlorophyceae.
prostrate	creeping along the substratum.
pseudocilium	flagellum-like structure found in the Tetrasporales.
pseudoraphe	axial area of a diatom valve without a raphe.
punctum	small hole.
pyrenoid	rounded protein body embedded in the chloroplast.
quadrate	square or rectangular.
raphe	deep longitudinal slit in diatom frustule connected with motility.
reticulate	broken up into a lattice or net.
rhizoid	colourless, hair-like filament serving as an attachment organ.
rhomboidal	rhombus-shaped.
rostrate	beaked.
scalariform	ladder-like.
semicell	symmetrical half of a desmid cell.
septum	cross partition dividing a cell into chambers.

sheath	mucilaginous layer external to cell wall in cyanobacteria.
side view (desmid)	Edge view of cell looking down through the median constriction.
sigmoid	S-shaped.
sinus	the deep depression dividing semicells of a desmid.
spermatium	non-motile male sex cell.
spheroidal (prolate)	solid figure generated by rotating an ellipse about its major axis.
stellate	radiating, star-shaped.
stigma	a) isolated pore in central area of a diatom valve b) eyespot or photoreceptor within the cytoplasm of flagellates.
stratum	layer.
symbiont	component of a symbiosis or close association between 2 or more organisms.
synonym	a systematic name superceded by another brought about by the rules of nomenclature.
terminal	the end of a cell or filament.
test	see theca.
thallus	plant body.
theca	urn- or pot-shaped structure enclosing a cell.
trichocyst	minute specialized pit containing a long retractable filament.
trichome	chain of cells (cyanobacteria).
trifid	three-pronged.
uniaxial	containing a single main axis.
uniseriate	a single row of cells.
valve view	view of a diatom valve face.
ventral	margin closest to the axis, the less convex margin of a dorsiventral cell.
verrucose	covered in warts.
vertical view	(desmid) view along the main axis passing through the apices and isthmus.
zoospore	motile asexual spore.
zoosporangium	cell producing zoospores.
zygospore	thick walled resting spore formed by the fusion of two gametes.

INDEX

239

242

243

244